The Ocean World of Jacques Cousteau

The White Caps

The Ocean World of Jacques Cousteau

The White Caps

WORLD PUBLISHING
TIMES MIRROR
NEW YORK

*Without trees, without grass, without land animals, the **antarctic landscape** is still more beautiful than any other setting on the face of the earth. In short intervals between two storms, the sun dazzles the eye, flashes the distant mountains with purple or gold, and lends the ghostly shapes of snow and ice a blue-white-green reality. The approach from the sea is unforgettable.*

Published by The World Publishing Company

Published simultaneously in Canada
by Nelson, Foster & Scott Ltd.

First Printing—1974

ISBN 0-529-05161-3
Library of Congress catalog card number: 73-20895

Printed in the United States of America

Project Director: Steven Schepp

Managing Editor: Richard C. Murphy

Assistant Managing Editor: Christine Names
Senior Editor: David Schulz
Assistant Editor: Joanne Cozzi

Art Director and Designer: Gail Ash

Assistant to the Art Director: Martina Franz
Illustrations Editor: Howard Koslow

Vice President, Production: Paul Constantini

Science Consultant: Dr. David Schwimmer

Creative Consultant: Milton Charles

Typography: Nu-Type Service, Inc.

WORLD PUBLISHING
TIMES MIRROR
NEW YORK

Contents

No traveler to southern oceans has ever failed to be impressed with the abundance of life IN THE AIR ABOVE THE ICE AND WATER (Chapter VI). The seabirds include the great albatrosses of sailors' legends, the ubiquitous petrels, and the multitude of gulls, shearwaters, prions, skuas, and cape pigeons that make their home far from man's land.   84

Although it may seem strange to visitors from more temperate climes, animals form a major part of the RICHES OF THE POLES (Chapter VII). The diversity of creatures is not so great as might be found elsewhere in the world, but in terms of shear numbers of organisms, the polar regions are unsurpassed. From the shrimp-like krill, which are so numerous that they turn the sea into a red soup, to the giant whales that feed on them, the sea is a storehouse of riches. On land, the riches become visual, from the sparse but colorful vegetation to the spectacular aurorae, borealis and australis.   96

Fish in polar waters must adapt to a way of life BELOW THE ICE (Chapter VIII). Conditions include not only the ice cover, but also variations in salinity, depending upon whether the ice is melting or forming, and water temperatures sometimes below freezing. Yet many different species, perhaps a hundred, have found these conditions bearable, including the ice fish, with its white blood coursing through a semi-transparent body.   106

Biological rules of development and evolution have been based upon observations in THE POLAR LABORATORY (Chapter IX). The type and distribution of animals, the size and shape of the body, the length of appendages all fit into general rules correlating heredity and environment.   114

One of the major accomplishments of man is to live IN HARMONY WITH THE ENVIRONMENT (Chapter IX). The Eskimoes achieved this in their arctic homeland with a way of life that extracted everything useful from the habitat, wasted nothing, and exploited nothing. But today's residents of the area—military, scientists, pioneers, and modern Eskimoes—have forgotten the lessons of the past.   122

The history of man in the polar regions has been one of exploration and exploitation, especially of the animal resources like whales and seals. Now man must decide whether he will change this, whether he wants TO PLUNDER OR TO MANAGE THE POLES (Chapter XI). Will the despoiling continue, or will the direction follow the spirit of accord that brought about the Antarctic Treaty, which preserved that continent for science?   132

Introduction: Balance and the World

Life in nature is, in essence, a struggle. A struggle for food, for space, for safety, for perpetuation of the species. But also, a struggle against adverse conditions—heat, cold, salinity, drought, wind, mud, and dust. The plants and trees in a forest compete for air, water, and soil as fiercely as wolves, hawks, or sharks fight to maintain the pattern of their existence. There are two sets of conditions—the scarcity of water and the rigor of cold—that, where they hold sway, have almost obliterated all forms of life. These "desert" realms may be yellow with sand and rock as the Mojave, Sahara, or Gobi, or white with ice as the polar caps of the arctic and the antarctic. In either case, they constitute the forbidding limits beyond which the provinces of death begin.

The North and South poles, the top and bottom of the earth, are as different as an ocean and a continent can be, but the waters bordering the lands are surprisingly similar. Because water needs a lot of calories to warm up, a lot of cold to cool off; because it is a much better heat conductor than air or earth; because water requires such huge quantities of heat to freeze or to evaporate, the ocean is the great climatic moderator of our planet. While the air temperature may reach $+136°$F. in the Sahara and drop to $-126°$F. in Antarctica, the temperature of the seas rarely exceeds $+80°$F. or drops to $+30°$F. Paradoxically, warm-blooded creatures, such as mammals and birds, that must maintain a fixed central body temperature, are better equipped to resist the extremes of temperature—thanks to fat, fur, and vascular controls—than cold-blooded animals, which would literally freeze if water temperature were to drop to as little as—say—$+25°$F. In both Arctic and Antarctic oceans the water temperature remains fairly close to life's edge, and some of the Antarctic fish are believed to discharge a sort of "antifreeze" protein into their blood to avoid being turned into blocks of ice.

Another major paradox of the polar seas is that, so close to universal death, marine life is several times more plentiful than in any of the temperate or tropical areas of the ocean! This is because organic matter, rained down to the bottom of the seas, is decomposed by bacteria into nutrient salts and carried to the polar zones by deep-ocean bottom currents and back to the surface by upwellings. In these frigid but nutrient-rich waters, all the sun's energy during the uninterrupted summer day is turned into vegetable plankton; crustaceans thrive on these glacial meadows and become abundant food for larger animals. Unfortunately, the prodigious tonnage of living organisms produced in both frigid seas is distributed among very few species, perhaps because the variety of ecological niches is so limited. Diurnal habits, for instance, would be useless in place where dawn and dusk may be six months apart. The polar pyramids of life are high and thin, extremely vulnerable, ready to collapse if seriously disturbed.

As a matter of fact, the delicate web of life in the boreal and austral seas has already been endangered, partly because in such remote reaches, the crimes of man remained without witness. In the arctic, the last true penguin (Auk) was killed in 1948; the polar bear, the narwhal, and the sea otter are endangered; the populations of belugas, walrus, arctic fox, and wolves have been decimated. In the antarctic, only 6 percent of the whale population is left;

the Ross seal is on the verge of extinction; millions of penguins have been slaughtered by whalers, and boiled to make oil. If some species of penguins are abundant today, others like the king penguin have almost disappeared. Knowing the destruction man has wrought in the past few decades, I was unable to reconstruct what the first polar explorers must have witnessed. I sat in an Eskimo umiak, in the middle of the Bering Strait, imagining the mass northward migrations of walrus 300 years ago, before they fell prey to frustrated whalers. And in the antarctic, I often flew in *Calypso*'s helicopter in search of the few remaining whales and thinly scattered seals, while, below, millions of the smaller penguins feasted on the unused whale food. The splendor of the poles today is but a dim image of what they were before human intrusion. In the north as in the south, the dazzling landscapes are already stained by the sinister silhouettes of oil prospecting derricks.

We have done so much harm to our planet already . . . now we want to extend our destruction to those areas which are least capable of accommodating change. How much longer before we make a truce with our world, so that the ends of the earth will remain what they are?

Jacques-Yves Cousteau

Chapter I. The Polar Settings

Six million squares miles of ice and cold. The ice is bright, but not crystal clear. Unending opaque whiteness curves off into the horizon in the polar regions of the earth.

The similarities of the northern and southern polar areas are obvious, but so are the differences between them. The northern polar region is a vast ocean bounded by continents and islands; the southern end of the globe is a continent, which is for the most part buried beneath tons of snow and ice. And the ice in the two areas is dissimilar. The North Pole is covered with slightly salty ice formed from the sea itself, while the continental ice at the South Pole is compacted freshwater snow, which has been fused into the world's longest single piece of ice.

Each area contributes to the continuous flow of ocean currents. The salt left behind as sea ice forms during the long polar winters in-

---

> "Each pole is a source of icebergs, those ghost ships that display only one-seventh of their mass above the surface of the water."

---

creases the density of the surrounding water. This heavier water sinks and the resulting convection stirs the water as deep as 165 feet. These changes of density generate movements and become an element of the deepwater circulation system in which there are several layers of water with varying thicknesses moving about in different directions around the globe. The stratification of water is based on temperature and salinity—which together determine density—and as these conditions change, the waters rise in upwelling or fall in downwelling in different parts of the world.

While polar waters eventually move about into the major oceans, polar ice has a different story. Each pole is a source of icebergs, those ghost ships that display only one-seventh of their mass above the surface of the water. Most of the arctic icebergs are calved on the west coast of Greenland; they then drift southward to menace the shipping lanes of the North Atlantic. Even more imposing are the southern icebergs, which present a flat-topped, sheer-cliff face to the world as they head north.

Icebergs can be spotted and observed but the polar regions themselve remain largely unknown. Many of the deep features and characteristics of polar ice are hidden. Nowhere does man interfere less with natural forces. At the poles the scouring, sculpting wind cuts eerie shapes amid the snowfields. The pressure and weight of the ice itself is responsible for many of the topographical features, such as the gaping crevasses, jagged peaks, and spectacular icefalls. The action of the sun on the surface produces meltwater, which cuts channels through the ice. The sun's heat helps produce névé, or firn, that substance intermediate between snow and ice. The sun also warms stones and rocks; the heated stones then begin to bore through the ice. Ice that melts and recrystalizes produces foliation in the glaciers and helps record the passing of time in much the same way growth rings of trees do.

Perhaps the most valuable feature of the cryosphere—the world of cold—is the extreme contrasts it offers to an inquiring man.

*At the **approach of spring in northern Alaska** the endless night begins a reversal toward six-month days. The warming sunlight melts the ice, which drips and refreezes, forming icicle daggers.*

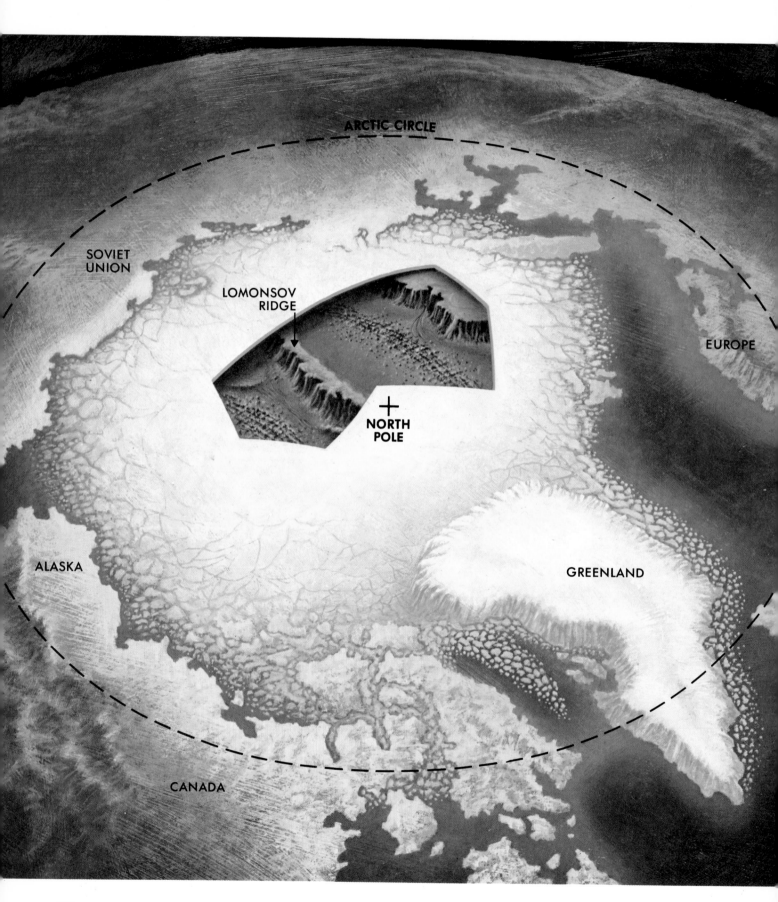

North of the Lands

The arctic world is a sea world—the north end of the globe is covered by an ocean dotted with masses of floating sea ice. There is no land to mark the spot of the North Pole. But surprisingly this region is not as cold as the interior of landmasses to the south (Alaska and Siberia)—even in the arctic the ocean tempers the weather.

Traditionally the arctic has been defined as that part of the globe above 66°33′03″ north latitude; it therefore includes sections of Canada, the United States, the Soviet Union, Finland, Sweden, and Norway and most of Greenland. But perhaps a better definition of arctic would be that northern area where no trees grow. Just as mountains have a tree line above which the climate is inhospitable to growing trees, so does the globe have such a timberline. In some places the line moves well above the Arctic Circle and skirts 75° north latitude, while in other places it approaches Cape Churchill in Canada, bypasses Nome, Alaska, and reaches as far as 56°, the latitude of Copenhagen, Denmark.

The arctic is hard to describe in other than general terms. It is marked by frequent high winds; long, cold winters and short, cool summers; low precipitation (lower than many desert areas of the world); and permafrost, the layers of subsoil that remain frozen throughout the year.

Water is the primary feature of the arctic landscape. Flying toward the North Pole in summer, one first sees frozen land and then the sea ice on its border. A few miles out, the dark ocean waters make their first appearance, gradually becoming lighter green studded with pieces of ice called brash, which are splinters of the mammoth floes. Beneath the sea is an ocean basin as warped and rifted as any on the face of the earth. The major feature is the submerged Lomonosov Ridge, which divides the arctic floor into the Amerasia Basin on the Pacific side and the Eurasia Basin on the Atlantic side.

The water flow of the Arctic Ocean is part of the worldwide oceanic system, with about 60 percent of the outflow spilling into the Atlantic between Spitsbergen and Greenland. Because of land barriers and the earth's rotation, there is almost no flow from the Arctic into the Pacific. However, about 35 percent of the Arctic's inflow comes from the Pacific. Most of the other inflow into the Arctic comes through the Norwegian Sea.

Unlike the antarctic region, the arctic has been explored and settled. The earliest explorers were sailors from Europe, interested in finding a shorter route to the Orient; not unexpectedly, their efforts were frustrated by the everpresent pack ice. The peoples that populated the subarctic region were primarily of Mongoloid stock—North American Eskimos, various Siberian tribesmen, and the European Lapps, who are believed to be partially Mongoloid. There may be something physiological in this racial stock which allows superior adaptation to cold, but there were still many cultural adaptations required. Homes had to be built with two entrances, providing a "heat lock" so that interiors would not be chilled with every coming and going. There was also the problem of "human fog," that condensation from human and animal exhalations and fires which hangs over a settlement. Until technology came non-Mongoloid peoples had to adopt a similar way of life.

*Lying under a thin cap of sea ice is the **Arctic Ocean**. It is surrounded by the landmasses of the Northern Hemisphere, and below it, shown in the cutaway, are rugged bottom features—notably the Lomonsov Ridge. The ice on Greenland and some of the northernmost land areas is many times thicker than the polar cap because it is underlain by land, which imparts less heat than the arctic waters.*

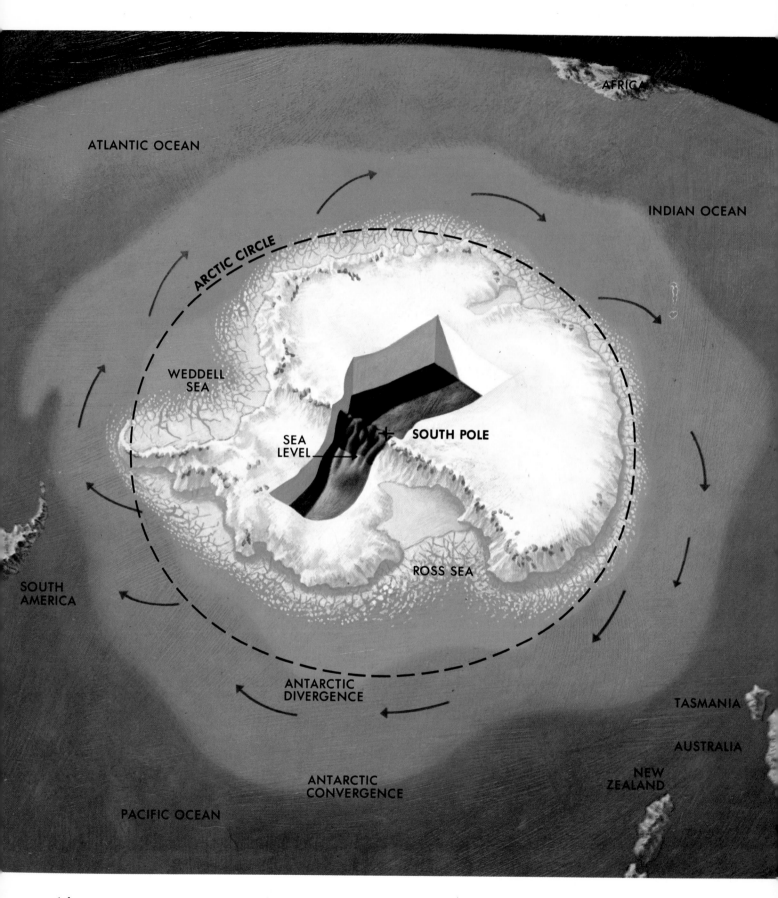

AFRICA

ATLANTIC OCEAN

INDIAN OCEAN

ARCTIC CIRCLE

WEDDELL SEA

SEA LEVEL

SOUTH POLE

ROSS SEA

SOUTH AMERICA

ANTARCTIC DIVERGENCE

TASMANIA

AUSTRALIA

ANTARCTIC CONVERGENCE

NEW ZEALAND

PACIFIC OCEAN

South of the Oceans

The Arctic is an ocean surrounded by continents, but Antarctica is a continent surrounded by ocean. Therefore the arctic climate is moderated by the sea, but the antarctic is subject to the extremes of continental weather. The lowest temperatures on earth are recorded in Antarctica—including one of − 126.5° F. The mean annual temperature can be as low as − 67° F.

The southernmost continent is roughly circular in shape. Palmer Peninsula extends up toward South America, well out of the Antarctic Circle, and is punctuated with mountains which are an extension of the Andes. Though ice and snow are the dominant features of the landscape, Antarctica has coastal cliffs that are scoured by wind and are too steep to retain snow. In some places, the sun of summer—November to March in the Southern Hemisphere—raises temperatures above freezing, and areas with thin snow cover become exposed. Scattered among these rugged coastal mountains are ice-free valleys, which are called dry valleys by the antarctic explorers.

Just as the borders of the arctic region are difficult to pinpoint, so are the limits of the antarctic. Astronomers have designated a theoretical line at 66°33′03″ south latitude as the Antarctic Circle. This line is determined by the rotation of the earth around the sun. The angle of its exposure to the sun's rays defines the antarctic—like its northern counterpart—as that area which has alternating six-month days and nights.

A better boundary might be the Antarctic Convergence, the point where the colder and saltier waters flowing north drop below the warmer waters of the Atlantic, Pacific, and Indian oceans. This boundary, which runs roughly between 60° and 50° south latitude, comes within 300 miles of Tierra del Fuego. It is an irregularly shaped circle that marks a sharp distinction in surface temperatures of the water and in the life-forms found in the upper reaches of the sea.

Perhaps the most spectacular frontier of the antarctic region is the almost impenetrable ring of ice, hundreds of miles wide, which girds the continent tightly in winter. In summer, when the solidarity is broken, the closely packed ice drifts northward until it is stopped by continental barriers and strong winds. But the ice still remains thick enough to prevent most surface ships from approaching Antarctica.

The antarctic landmass is the continent that is least hospitable to life. It is still frozen in the ice age and locked in isolation. Only within the last century has man set foot on the continent that was known for centuries as *Terra Australis Nondom Incognita*—unknown land to the south. Much is still unknown about Antarctica. That it influences weather throughout the world is accepted, but there is still speculation on how much and why. We know there is land beneath most of that ice, but its extent and topography are still being determined. It appears that there may be a land area, East Antarctica, and a series of islands in the west, similar to the arctic islands of Canada.

*Surrounded by the only waters that completely encircle the earth lies the continent of **Antarctica**. Scientists disagree about the name for the ocean that is formed by the circling waters—some prefer "Southern Ocean," some "Antarctic Ocean"—but except for the tip of South America, there are nearly 1000 miles of open water separating Antarctica from the nearest landmass. The Antarctic Circle is a theoretical boundary for the area, while the Convergence and Divergence represent physical and biological limits for the waters. The cutaway shows the incredible thickness of antarctic ice and the effect this enormous weight has had in pressing some land surfaces of Antarctica well below sea level.*

Ice from the Sea

The most extensive ice area in the Northern Hemisphere is the arctic ice pack, which is a floating slab of frozen seawater. The pack expands and contracts with seasonal changes, but even in August it extends almost from Point Barrow, Alaska, to very near the coast of Siberia, Greenland, and the Canadian Archipelago. Sea ice, of course, also occurs in the antarctic, but it is thinner than its northern counterpart.

In winter the arctic ice pack extends downward and averages between 8 and 12 feet in thickness. Further growth is inhibited because the covering ice, being a poor conductor, prevents the heat of the water from being absorbed by the atmosphere. From August to March the sea ice doubles in size. The new ice is called annual, or winter, ice and is distinguished from the polar ice that lasts from season to season.

As expansive and imposing as the ice pack is, it is not a stable formation. The pack is in almost perpetual motion, twisting, turning, and breaking into smaller pieces. Through this grinding action, the rough edges of the pieces are smoothed, and the result is often small, rounded chunks of ice that resemble pancakes 20 to 40 inches in diameter. In colder weather this pancake ice coalesces once again, forming sheets of scaly ice. Ice forming on especially calm bodies of salt water will sometimes take the shape of pancake ice without requiring the grinding action of the pack ice. The surface of the pack ice is often distorted as a result of ridges formed by the grinding pressure. Hummocks and cracks appear as a result of the pushing and pulling of winds, swell, currents, and tides. With the formation of hummocks, the action is accelerated since they present a saillike surface to catch the wind and are moved about more rapidly.

When the sea ice breaks into pieces, the smaller units—which may measure miles across—are called floes. Cracks that are large enough to steer a ship through are called leads, and those openings that are completely surrounded by ice are called polynyas, from the Russian for hollow or open. Both leads and polynyas are formed by the grinding action of the pack ice and floes. The amount of open water in the arctic ranges from 5 to 8 percent in the winter to as much as 15 percent in the summer.

Sea ice is a relatively complex material, and its physical properties are dependent on temperature, salt content, crystal structure, and air bubbles. Salt has almost no solubility in

solid ice and most of the salts are trapped in pockets of liquid brine. The lower the air temperature at the time the sea ice is formed, the higher the salinity of the ice. But as the temperature of the ice increases, and especially as it approaches the melting point, the brine escapes from the ice, greatly reducing the salinity. This migration of salt out of the sea ice is so effective that Eskimos have used sea ice as a source of drinking water.

Jostling in the sea, according to most scientists, produces a rounded pancake shape (above and opposite) in some types of young sea ice. This "pancake ice" also has turned-up edges from the impact with other pieces. Calypso crewmembers observed pancake ice forming within a half-hour's time after the water's surface began to freeze. This makes one wonder whether we truly understand all the subtleties that govern the normally slow freezing of water.

During spring months, sea ice (below) begins to crack, forming open water stretches called leads, the only navigable passages in coastal polar waters.

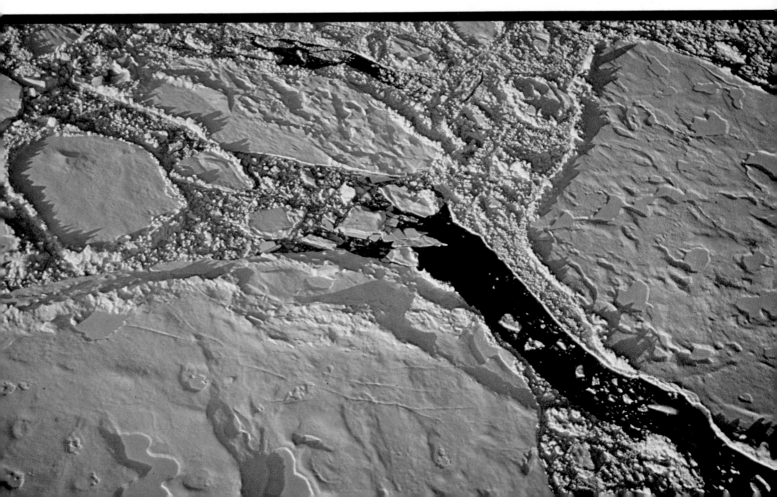

Freshwater Ice

It has been calculated that the ice mass covering Antarctica contains nearly 90 percent of all the snow and ice in the world. This freshwater mass is the result of 20,000 years of snow accumulations. The transformation of snow to ice requires the interaction of time, wind, and sun and the pressure of layers of snow, which forces out the air between the snow particles and compacts the crystals. This process continues for scores of years until a material is formed that is neither snow nor ice, but something in between called firn, or névé. This substance is denser than snow since the ice particles are beginning to become connected. With additional time and increasing weight of overlying layers, firn will become ice. Of course, temperature is one of the most important factors in this conversion—summer temperatures must not be high enough to melt the winter snow.

This mass of antarctic ice is a glacier that averages more than a mile in thickness and measures 2500 miles in breadth. The glacier is not a flat tabletop; in addition to the sculpting action of the wind and the irregular melting by sun's rays, the ice has a tendency to creep. Solid matter in large amounts, whether it is ice in a glacier or rocks beneath the earth's crust, displays a plasticity usually associated with fluids. The substance remains hard and brittle if subjected to sudden force; but if constant, yet not extreme, pressure is exerted, it will flow, however slowly. For this reason, the glacier covering Antarctica flows toward the edges of the continent, along the contours determined by the topography and the pull of gravity.

Another form of solid freshwater is the ice island, found among the saltwater floes of the Arctic Ocean. Ice islands are rare but are easily detected because of their extraordinary thickness. An ice floe is only as thick as the pack ice—8 to 12 feet—but an ice island can be 200 feet thick. It is a chunk of ice broken off a glacier perhaps on Greenland or more likely on Ellesmere Island. Because of their thickness, ice islands are much more stable than floes and are less likely to be crushed and cracked in the roiling ice pack of the Arctic Ocean.

*Every snowflake is different and forms a variation of the hexagon. The **polar freshwater ice** is made up of uncountable numbers of snowflakes compacted to the density of ice.*

Floating Ice Mountains

Icebergs come in two styles—northern and southern. The northern, or arctic, iceberg is a jagged peak of irregularly shaped ice broken from a land-based glacier; the southern, or antarctic, iceberg is a flat-topped chunk calved by the massive ice shelf surrounding Antarctica.

The common northern iceberg is spawned, or calved as glaciologists say, when the glacial ice of Greenland reaches the waters of Baffin Bay, or the Greenland Sea. Without terra firma to rest on, the ice mass is subjected to the alternating flex of the swells and tides and pieces are broken off. As many as 10,000 icebergs may be calved in a year's time, some of which can tower 500 feet above the surface, with 85 percent of the mass still submerged. Currents and wind push the icebergs into the North Atlantic where they can drift into the Labrador Current toward the Gulf Stream and present a real hazard to oceanic transportation. An arctic iceberg was responsible for one of the

*The flat, tabular look of a **southern iceberg** (below) reveals its origin as an ice shelf, most likely the massive Ross Ice Shelf, or perhaps the Weddell Ice Shelf on the opposite side of Antarctica.*

*The mountain-like peaks of a **northern iceberg** (above) can often be seen in the Greenland Sea. The icebergs that calve off the island's glacial ice cap menace ships in the Atlantic.*

most notable maritime disasters of modern times. It occurred in 1912 when the supposedly unsinkable *Titanic* went down in the Atlantic, taking the lives of 1500 people.

As these icebergs drift into warmer waters, as far south as Bermuda and the Azores, they

begin to melt and deposit silt, pebbles, rocks, and large boulders on the ocean floor. This terrestrial material was picked up during the thousands of years it took the glacier to move across the face of the land, scraping and scouring the earth.

The tabular icebergs which are found in waters of the Southern Hemisphere have no counterpart in arctic waters. These table-topped formations measure up to 100 miles across and have faces like sheer cliffs. The antarctic ice shelf, unlike the northern ice pack, is not frozen seawater, but is an extension of the freshwater glacial icecap. As a result, it is a more permanent feature and grows several times thicker than the sea ice of the arctic. Antarctic icebergs may rise as much as 250 feet above sea level and have been sighted as far north as 26° south latitude, nearly the latitude of Sao Paulo, Brazil. Southern icebergs, though, are very much larger than the ice floes found in the Arctic Ocean and would smash to brash any sea ice they encountered.

Ice Features

The vast expanse of the polar regions might at first glance seem to be an unbroken landscape of ice and snow, but there are many features that are coldly beautiful, treacherously dangerous, or startlingly unique. The ice above the water, the pack ice, is subjected to great horizontal stress, and as a result, pressure ridges are formed in the center of ice floes. If this horizontal force is extremely strong, floes may overlap and be piled one on top of another. Occasionally a piece of ice may be frozen in an upturned position, providing a rare formation that offers shelter or a hiding place for a hunter.

The ice shelf of the antarctic, because of its extraordinary thickness, is much less affected by pressure than the pack ice. As the glacial icecap pushes toward the sea, it continues to cover the water as it did the land. In protected areas like the Ross Sea, the glacial ice flattens because of the annual accumulation of about 20 inches of snow and the melting effect of the relatively warm water acting on the bottom of the shelf. The floating, raftlike shelf—which is anchored on land—is over 2000 feet thick in some places and is 500 miles across in its widest sections. The Ross Ice Shelf has a surface area of 160,000 square miles, which makes it nearly as large as continental France.

The ice covering the land also has distinct features, although in general terms the hills and dales are less extreme and smoother than those of a continental landscape. Probably the most unusual, and most dangerous, formations are the crevasses. These are open cracks or fissures in the surface of the ice caused by stresses when passing over a raised feature, such as a hill. Crevasses, which have steeply sloped sides, may be several yards wide at the top and 150 feet deep at the bottom of the wedge. There are mammoth

Crevasses (above) cut the snowscape like knife wounds, sometimes opening gaps large enough to trap a man like this one on the coast of Greenland.

*Drivers of **snow vehicles** (below) must be constantly on the lookout for crevasses to avoid sudden pitfalls or more serious disasters.*

crevasses in Antarctica that are 100 yards across at the mouth, 200 feet deep, and several miles long. A glacier that is severely crevassed and broken, because it has passed over a cliff or down a steep slope, produces an icefall. A jagged peak or pinnacle, called a serac, forms when crevasses intersect.

Crevasses are often hidden by snowbridges, which are formed when wind-sculptured snow cornices on either side of the crevasse are joined in the middle by additional snowfall. With the formation of an ice cover and more snow, these bridges may become strong enough to support a man and a dogsled, but usually not heavy vehicles like Sno-Cats. Motor vehicles used in the Antarctic usually have an electrical crevasse detector, which is pushed 30 to 40 feet ahead of it. The detector's saucer-shaped discs send electrical impulses into the snow, creating an electrical field. If the material beneath the snow changes, as it would in a crevasse, the instruments measuring the electrical field record this and activate a warning signal.

The fierce winds which form the snow cornices and snowbridges also carve natural formations called sastrugi, from the Russian word for groove. Sastrugi are wavelike ridges of peaked and pitted snow, formed by the wind whipping across a level surface.

Wind may also hollow out depressions in the snow, but it is unlikely that this is how snow caves are formed. More likely, ice-lined caves are carved by meltwater, which scoops out a channel that eventually acquires a snow roof in the same way snowbridges cover crevasses. When they can find snow caves, polar bears in the arctic use them as dens.

*Meltwaters can seep through cracks and small channels in the ice, widening them into long **tunnels and caves.** In some instances, the water-level marks can be seen as ledges lining the sides of the cave.*

Chapter II. Coping with Cold

However hard cold is to define, it is a physical fact of life, and it is a common characteristic of both polar regions. Some say that cold begins when a body starts to shiver, or when water starts to freeze. Time is a factor, when prolonged periods of low temperatures are used in the definition. And both wind and humidity increase the effects of cold by exaggerating the chilling properties.

The South Pole has an annual average temperature of −50° F. In such a climate, the only animals able to survive are those endowed with the physiological flexibility necessary to adapt to extreme cold. Temperature alone is critical for cold-blooded creatures when their body fluids are about to freeze. Warm-blooded animals, on the contrary, have to be protected against *heat loss,* which depends on many factors other than temperature. The still, clean, vast, mysterious, beautiful reaches that make up the polar

---

> "This radical alternation of day and night has little or no effect on the animals now living in the region."

---

regions harbor a limited number of species, however plentiful their numbers may be. The overhunted polar bear, for instance, is able to withstand the constantly freezing temperatures of the ice pack around the Arctic Circle. Until man came, the polar bear had no terrestrial living enemy and was able to cope successfully with the challenges of an otherwise harsh environment.

Seals, warm-blooded marine mammals, have developed layers of fat to protect them from the chilling effects of sea water often slightly colder than 32°. The escape into water only partially protects northern seals from land-living polar bears because they must return to breathing holes where the bears may be waiting for them. For safety, seals maintain a network of such breathing holes throughout the winter months, when the ocean ice cover reaches several feet of thickness. While the air-breathing seals have coped with polar living by developing insulating layers of fat, all the terrestrial creatures have thick layers of fur which keep them warm enough so that an increase in body metabolism is unnecessary except in extreme circumstances.

One of the more bizarre features of the polar area is the midnight sun, that oddity of nature which causes the sun to shine continuously for six months of the year and then hide for the next six months. This radical alternation of night and day has little or no effect on the animals now living in the region. What might have happened, however, is that creatures with strongly diurnal habits, such as daytime feeding and nighttime sleeping, never migrated to polar areas because of the absence of a 24-hour cycle.

All the other difficulties of living in the polar regions stem directly or indirectly from the cold. The animals must deal with ice and snow covering the land in the great north or the sparse vegetation in the lands of the subarctic. Food is difficult to come by on land and must continuously be sought for. Underwater, on the contrary, food is abundant. Cold is a barrier to life, but at the same time it serves as protection; predators are fewer than in warm tropical areas, and competition for life and for space is less intense.

In the polar regions, the cold is the prime mover of all life processes. To survive it one must dress carefully. Moisture in his exhalations give this **polar visitor** *a chinful of ice.*

Highs and Lows

Statistics, as has often been noted, can be used to prove almost anything. For example, two places that have a daily mean temperature of 50° may have very different climates. In one spot temperature during the day may be 60° and during the night 40°; but the temperature of the other location may range from 100° to 0° in a 24-hour period. Usually, though, mean daily temperatures give a general indication of an area's climate.

Some locations, such as deserts, display a wide daily variation in temperature; because they are dry, they heat up very fast when the sun is shining and cool off equally quickly during the night. On an annual basis, however, the variations may not be all that great. In other places, such as coastal regions, the difference between daily highs and lows may be relatively small, while there are great variations between daily mean temperatures during summer and winter.

The fluctuation of air temperatures between night and day in continental locations is closely associated with the temperature of land surfaces and is affected by such things as precipitation and cloud cover. An overcast sky can prevent solar radiation from reaching the earth, while a clear sky at night allows heat to escape high into the atmosphere. As a general rule, because of the very high caloric capacity of water, there are smaller variations in oceans and coastal areas, and greater diurnal ranges in the continental interiors. Other factors influencing daily variations are the direction of the prevailing winds—whether from the land or the sea—and atmospheric disturbances.

Like the daily variation of air temperatures, the annual march, or range, is affected by proximity to oceans which act as calorific flywheels. In the Northern Hemisphere areas that are strongly influenced by the sea, February is usually the coldest month and August the warmest, while deeper into the continents, January is often the coldest month and July the warmest.

Lines called isotherms are used to depict graphically temperatures throughout the world. *Isos* is the Greek word for equal and thermal comes from the Greek word for heat. These lines, which connect all points on the map having the same temperature for any given time or period, generally run east-to-west, showing that there is a correlation between temperature and latitude. Coastal

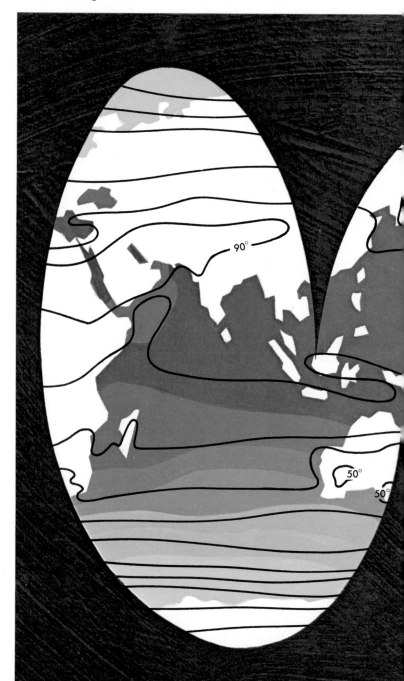

areas with strong ocean currents and mountainous altitudes bend these isotherms accordingly. In some places the isotherms are close together, signifying a steep temperature gradient, such as occurs in polar regions during the winter months.

Because of the concentration of landmasses in the Northern Hemisphere, the mean annual temperatures are slightly higher north of the equator than south of it. Whether this is due to the presence of Antarctica or, is the cause of the extreme cold at the South Pole is still not fully determined.

The extremes of temperatures cover an almost unbelievable range of more than 250° F. Readings above 130° have been recorded in the Sahara Desert in Libya; at the same time of the year, when it is winter in the Southern Hemisphere, temperatures have dipped below minus 120° in Antarctica.

*On an "orange-peel" projection of the globe are shown the average **water and air temperatures** for the month of July. The lines (isotherms) show that the southern region is in winter, and that the weather bands are almost even due to the unbroken antarctic ocean circulation. The water temperatures are less affected by seasonal differences.*

TEMP. °F	
86+	
77-86	
68-77	
59-68	
50-59	
41-50	
32-41	

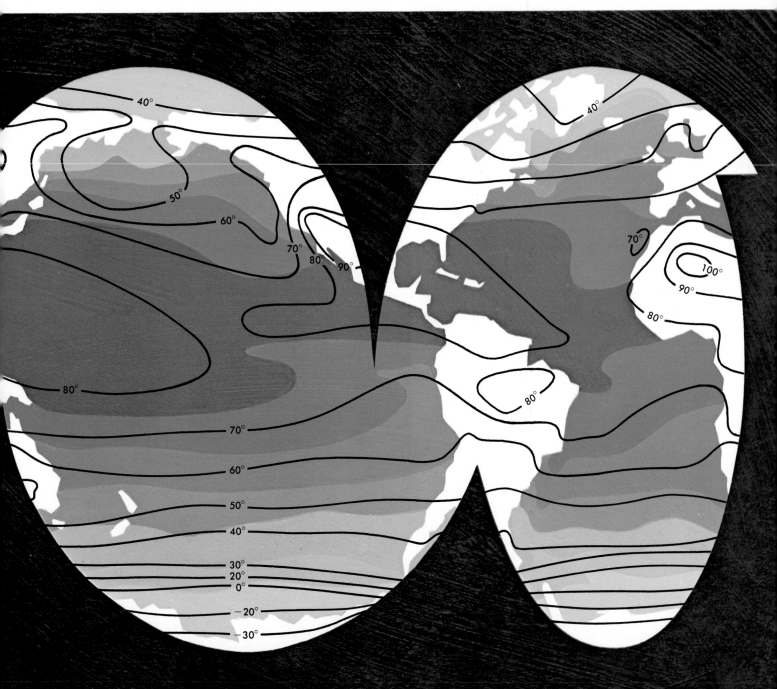

Angles and Degrees

The sun is the source of life on this planet. Without solar energy there would be no photosynthesis, and without photosynthesis there would be no plants. And without plants animals could not sustain life. All energy in the world comes from the sun.

The sun's rays reach the earth in a flow of short wavelengths, of which about 20 percent are absorbed by the atmosphere directly. The rest pass through the atmosphere and strike the earth itself, where the longer waves are converted into heat and radiated back out toward space. The atmosphere, through a complicated process of convection, evaporation, and layering, traps much of this radiated heat which helps determine the earth's mean temperature of $60°$ F. Without the atmosphere, the mean earth temperature would probably be about $15°$. Water has a greater heat-absorbing ability than the soil and rocks on land, so the oceans—which cover 70 percent of the earth's surface—generally determine global temperatures. For this reason, there is much less of a temperature range in the oceans than there is in the continents. And land areas bordering oceans are generally more temperate and less subjected to wide fluctuations of climate than the interiors of continents. The most extreme temperature variations are found in continental areas that are behind mountain ranges that block the moderating winds blowing from the oceans. The absorption and radiation of solar rays are closely connected with the shape and position of the earth vis-à-vis the sun. The equator and tropics receive the sun's rays directly at an almost $90°$ angle. As a result, in these areas both the earth and the atmosphere gain more heat than they lose. Ocean currents and tropical storms transfer the excess heat to higher latitudes. At the poles, however, solar radiation strikes the earth at a much sharper angle, and the net effect is that less incoming radiation is absorbed than there is outgoing radiation lost.

*The **higher latitudes** denote the polar regions, and the scenery of the area is stark and icy compared to that of the tropics or temperate zones.*

Because of the complex interchanges of heat between the ocean, atmosphere, and land neither are the poles growing colder nor equatorial regions becoming warmer; taking the earth as a whole, the heat gains equal the loss through radiation. The indispensable heat transfers from the warmer to the colder areas occur mainly through the atmospheric and oceanic circulation systems.

*Lines of latitude mark the north-south positions on the earth. In the **lower latitudes,** the scenery at sea level is lush and tropical like that of Moorea.*

*The illustration (above) shows the longest summer day (24 hours) at the North Pole; on that day the antarctic receives no sunlight. The earth is suspended at a 23.5° angle with respect to the sun, which explains **why the polar regions are cold.** Sunlight hits the poles at a very oblique angle, and it therefore is diffuse and weak. The earth's rotation on its axis follows an eccentric (off-center) path which takes almost 26,000 years for each cycle. Superimposed on this eccentricity (called precession) is a smaller wobble (nutation) caused by the alternate pulls of the sun and moon. Nutation is too small to illustrate on the scale here.*

Long Days and Nights

The phenomenon of the midnight sun is a longstanding symbol of the polar regions. The same factors that make the midnight sun shine also cause the cold, harsh climate. The six months of winter in which the sun is out of sight at each pole are the result of both the earth's tilt on its axis and its orbital path in its annual revolution around the sun. At one point the earth offers one pole to the sun, and six months later—one-half of an orbit—the other pole is facing the sun. Thus in the arctic and antarctic the amount of sunlight varies from 24 hours at the summer solstice to zero hours at the winter solstice.

This six-month variation means that for half the year, one or the other pole receives no solar radiation whatsover. Even in the six months when the sun's rays reach each pole, the angle is so oblique that there is a net loss of heat. Part of this is due to the reflecting ability of ice and snow.

Given the low amount of solar radiation reaching the poles, it is not surprising that their dominant feature is the rigor of the climate. What is more surprising is the lack of precipitation, for not only do the highest latitudes have the lowest summer temperatures and lowest mean annual temperatures, but they also average less than 10 inches of precipitation annually. Because of the cold, however, there is also a low rate of evaporation, there is usually sufficient moisture to support life in those areas where temperatures of the soil rise above freezing.

The types of climates in the polar regions include icecap climate, where the temperature never averages above freezing; tundra climate, where the average temperature is above 32°F. during at least one month of the year; and the subarctic or boreal climate, where the temperature may average 50° in the warmest month.

Fur and Fat

Most animals find it diffcult to cope either with cold or with great ranges in temperature. Warm-blooded animals, however, have adapted well to living in areas where the average temperature range can be as much as 60°F. annually, with even greater extremes at times. Mammals, including man, and birds have developed various means of conserving their high and constant central temperature, in spite of terrifying cold weather conditions. The alternate method of fighting cold is to increase metabolism so that more body heat is generated. This may be useful in tropical and temperate zones, but it is insufficient in polar regions partly because of the great amount of food that must be consumed to achieve this. Food is so scarce on the land of the polar regions that animals would have to spend too much time and energy hunting for food.

*The thick layer of blubber found in the **Weddell seals** and many other polar animals has enabled them to thrive in cold conditions that naked man would not survive more than minutes.*

Instead, land-based polar creatures have generally protected themselves by developing insulating methods—either increased body fat or thick layers of fur, or both. In addition, some animals have an internal system that conserves heat by not allowing it to be dissipated through the extremities. The parts of the body outside the central core—such as legs and snouts—can withstand very low temperatures without being damaged. In some cases, these extremities can even be cooled to very near the temperature of the air, while at the same time remaining completely functional.

More common, however, is the use of blubber and hair for protection against loss of body heat. The effectiveness of fat as an insulating material has to do with its melting point. In the extremities of the animals, fat has a very low melting point. Eskimos use fat from caribou's feet as a lubricating agent, because it remains soft even at temperatures as low as freezing. By contrast, internal fats have a much higher melting point. Marrow fat in the bones of a caribou is solid at 65°. Aquatic animals make the most effective use of blubber as insulating material.

Nonaquatic polar creatures combine body fat with layers of hair. The insulating prop-

The musk ox relies on a thick undercoat and a long overcoat of fur for survival in the cold. To weather the worst storms the herd forms a ring, with the young and females in the center.

erties of fur are well known and are so effective that the arctic fox is unaffected by temperatures as low as −30°F. Some land animals, such as the musk ox, develop great amounts of hair as protection, but this method of insulation is effective only in relatively large animals. Most smaller creatures are absent from the arctic. Lemmings are the smallest mammals found above the Arctic Circle, although the fox and hare range further north. Generally, even in the milder subarctic climates, smaller animals generally burrow in and hibernate for the coldest part of the year. The musk ox, however, is so well protected from the cold by its thick coat of hair that even during the most fearsome blizzard it need not seek protection. Rather, the small band gathers in a wedge-shaped or diamond-shaped formation called a yard. They are so closely packed that they sometimes step on each other's hooves. The less hairy young animals and the females are inside the guardian ring and the older bulls are on the outside. This group will wait out a storm for days, if need be, and not show any ill effects.

Divers from the Calypso *use holes in the sea ice to* **explore polar waters.** *When they emerge, there is usually food, drink, and shelter awaiting them.*

Food and Air

Adapting to the polar environment means more than just developing ways of staying warm. It also means having to deal with constant ice and snow, rough-blowing winds, cold waters, and relatively sparse food supply, particularly for land dwellers.

The subcutaneous layer of fat, common to so many polar animals, not only protects the animal from the cold but also functions as a food reserve in difficult times. Finding food is a problem for all land-based mammals in the arctic environment. Polar bears will wait for hours beside a seal's breathing hole, waiting for a potential meal to surface. Wolves have been observed devouring an injured

member of their own pack. And extreme hunger has driven musk oxen to seek food from man if they should happen across a rare human settlement.

Seals and other marine creatures of the polar regions have fewer problems obtaining food, for the cold oceans are surprisingly rich in life. But seals, being air-breathing aquatic animals, must cope with the ice cover over the water. In the summer months the ice is broken up enough so that open water is no problem, especially considering that seals are capable of swimming as far as eight miles underwater while looking for a spot to surface. Even if there is some ice on the surface, seals have little problem breaking through a layer four inches thick. They simply butt the

*Eskimo hunters and hungry polar bears often wait patiently near a **seal's breathing hole,** knowing that the seal must emerge periodically for air.*

ice with their heads to form a breathing hole. In the winter months, however, seals must cease their roaming and remain in a relatively small area, keeping open a number of breathing holes by repeatedly breaking the ice that forms over them. Where head butting is insufficient, the seal uses its teeth to gnaw at the ice, enlarging the hole so that at least its nose can protrude above the water surface to breath. In ice several feet thick, the hole may be very wide at the waterline but tapers to an opening of only a couple of inches in diameter at the surface. As the seal continues to keep the hole open, bits of ice are pushed above the surface, forming a mound. These mounds may acquire a snow covering, but enough air gets through to

serve the seal's purpose. In the arctic, however, these mounds signal the presence of breathing holes, and both Eskimos and polar bears are alerted to a seal's territory. They may have to wait several hours for the seal to use a particular hole, because a seal usually maintains several openings in an area. A mother polar bear and her cub may work in tandem, with one guarding a specific hole and the other going around plugging up all the other breathing holes in the vicinity. Sooner or later the seal will surface and the polar bear—or Eskimo hunter who uses the same trick—will strike. In the antarctic, where seals have no natural land-based enemies, the holes are kept much larger so that the seals can use them to feed in the sea.

Poles of Cold

There are several words used to describe the subarctic environment, from the common tundra to the Indian muskeg to the Russian taiga. But whatever their names, such areas are only slightly less inhospitable than the polar region itself. There are several different types of habitats—tundra is the broad, treeless planes of the northlands; the Canadian muskeg is a moss-covered bog or marsh composed of thick layers of decaying organic material; and the taiga of Siberia is covered with well-spaced conifers.

Like the climate inside the Arctic Circle, the subarctic weather is dominated by cold. German climatologist Julius Hann once related his experience in the Siberian taiga this way: "It is not possible to describe the terrible cold one has to endure; one has to experience it to appreciate it. The mercury freezes solid and can be cut and hammered like lead; iron becomes brittle and the hatchet breaks like glass; wood, depending upon the degree of moisture in it, becomes harder than iron and withstands the ax so that only completely dry wood can be split."

Animals who choose, or are forced, to live under these severe conditions are rare, for only a very few species have taken up permanent residence there. They range from the tiny lemming and arctic hare to foxes and wolves and up to the larger caribou, musk ox, and polar bear.

The cold of the region inhibits the growth of plant life, so that herbivores and grazers must spend much of their time searching for food. Massive herds of caribou and smaller bands of musk oxen are constantly on the move hunting for food. They are closely trailed by the carnivores, waiting for injured or stranded animals to prey upon.

*Deep soil layers of **permafrost** give a flood aspect (above) to tundra that is rugged and parched.*

When the animals of the subarctic lands aren't struggling with harsh winter conditions, they must put up with the peculiarities of continental climates which includes high winds and damp, foggy weather with severe storms in the spring and fall. The temperatures may range very much above freezing. At Yakutsk, Siberia, for example, the temperatures in July have averaged as high as 66° F. with occasional readings in the 90s. And Yakutsk is only about 300 miles from the Arctic Circle. The earth's thermal equator, the line of highest mean annual temperature, is located slightly above the geographic equator at about 10° N latitude. There are also poles of cold. In the North this is in the Yakutsk region around 140° longitude, while in Antarctica it is due east of the geographical South Pole, roughly between the Soviet stations at Vostok and the one in the "area of inaccessibility."

*Tundra animals like the **caribou** (left) wander in herds, looking for sparse vegetation to feed on.*

37

Chapter III. Icing Over the Past

So long isolated, Antarctica has been grudgingly yielding clues to its past. Understanding of the recent history of Antarctica requires a thorough knowledge of the mechanics of ice and snow on a large scale—that is, glaciers. Despite the presence of such massive ice formations throughout history, very little had been written or said about them until the middle of the last century when the Swiss naturalist Louis Agassiz became interested in the Alpine glaciers. It was then that scientific methods were applied to the study of these ice masses.

The mechanics of glaciers—their movement, correlation with climatic changes, sculpting and scouring ability—were all gradually revealed. But soon it became apparent that all the ice that had at one time or another covered the continents had not

---

**"Antarctica was at one time
a warmer, sunny place
with vast amounts of plant life."**

---

marched down from the mountains—some had come from the direction of the North Pole. The techniques and experiences of alpine glaciologists were applied to the study of continental glaciers. In a short while, it became obvious that several times in the past glaciers spawned in the arctic headed south in many directions. The late discovery of Antarctica and its relative inaccessibility and forbidding cold limited the study of its glaciation, which was an ice cap similar to but larger than Greenland's.

Study was then concentrated on the forces that produced the spread of glaciers and those that induced their retreat. Any great buildup of ice would necessarily lower the level of the water in the oceans and, conversely, any great melting of glaciers and ice caps would raise the sea level, causing great flooding in coastal and low-lying areas of the continents. In 1964, J. Tuzo Wilson theorized that periodically, when the accumulation of ice on Antarctica became large and heavy enough, the entire cap broke and slid into the ocean in a catastrophic manner that submerged the continents and sent enough ice to subtropical zones to cool the entire globe substantially, triggering the big glaciations. This hypothesis is very controversial.

To this day there is no satisfactory answer to the question: What causes glaciers and ice ages? The causes may be unknown, but many of their effects are evident, including the changes of landscape, the extinction of animals adapted to cold-weather living, and the development of other species to take their places. Searching for answers, scientists spent much time digging in the few accessible land reaches of Antarctica, mainly the Horlick Mountains and Queen Alexandra Range. Much to their surprise, they uncovered beds of glacial debris, which indicated that the continent had been covered with a glacier before acquiring its present ice sheet. Also uncovered, and even more surprising, were enormous coal beds containing fernlike fossils that indicated Antarctica at one time was a warmer, sunny place capable of supporting vast amounts of plant life. The stratification of these coal beds was correlated with similar stratifications on other continents—South America, Africa, India, and Australia. Somehow, at some time, apparently the continent of Antarctica was not as isolated as it now is.

*Flowing at a steady but slow rate, a **glacier comes down a valley,** widening and scouring the bedrock while simultaneously depositing rubble at its edges.*

Ice Flows

Glaciers come in all sorts of shapes and styles, but the most important ones in terms of the history of the earth are continental glaciers, such as those that now cover Greenland and Antarctica. Glaciers are formed of compacted snow whose formation changes from delicate hexagonal flakes to more rounded grains under the pressure of overlying layers of more snow. Once glaciers are formed, and if they are continually fed by additional snowfall, they begin to expand laterally either by being pulled down slopes by gravity or flowing slowly, demonstrating the phenomena of "creep," which is common to massive amounts of solid materials. Creep occurs as a result of the glacier's weight, even though its surfaces are brittle.

Continental glaciers hold fantastic amounts of ice; they cover several hundred square miles of surface and may be several miles thick. Even today, in the period of a receding ice age, the ice cover on Antarctica contains 90 percent of the world's ice and 75 percent of its fresh water. The enormous size and

weight of these glaciers submerges the land on which they rest to depths below sea level. This is true of both Antarctica and Greenland, where the surface of the land—as measured by seismic waves—is well below sea level in some places.

The most notable action of glaciers is their sculpting of land surfaces as they pick up rocks, both large and small, and scour the earth. The layer of rocks and boulders that a glacier carries cuts deeply into the existing bedrock, causing striations; the smaller particles in the glacier act like emery cloth and polish the rocks that the glaciers ride on. Continental glaciers can alter a landscape to a considerable extent, not only by their

Glaciers carve many strange shapes in the lands and mountains that they cross. U-shaped valleys (below) are easily recognizable, as are long, carved ridges called arêtes. Nunataks (opposite) are formed when ice flows through mountains, isolating individual peaks, such as these on Greenland.

sculpting effect but also by the deposits that they leave behind.

Since most of the debris a glacier carries is lodged on its sides and at its base—the surfaces that come in contact with the native ground—locations of past glaciers can be determined by the till and drift they leave behind. Till is the deposit of unsorted rocks left by glaciers, while drift is made up of particles carried by meltwater and laid down in more orderly fashion in the manner of most waterborne sediments. Perhaps the most obvious effects of glaciers are moraines, those ridges of soil and rock which the glacier plowed up while advancing and which it leaves behind, showing the location and points of furthest progress.

While the mechanics of glacial flow and deposit are well known, the precise cause of the damp, cool climates that foster glaciers is still not fully understood.

Tongues of Ice

In addition to the ice sheets covering Greenland and Antarctica, there are several glacial formations existent in the world today in mountain ranges. Some are only 35 miles from the equator, but at an altitude of 6500 feet. Glaciers are most common in the low temperatures and high altitudes of the Alps, Himalayas, Andes, and Rocky Mountains. Such glaciers are referred to as valley or alpine glaciers—to distinguish them from the more massive ice caps. Even though ice is a true rock formation, albeit with a very low melting point, it reacts much the same way that running water does—a property many crystalline solids exhibit under pressure or at temperatures near their melting point.

Just as continental glaciers carve the lands they move over, valley glaciers determine the character of mountains that hold them. Among the land formations associated with mountain glaciers are cirques, which are bowl-shaped depressions hollowed by the glaciers and from which the glacier flows. Often, after a glacier disappears, these fill with water to form a cirque lake or tarn.

Many of the terms used to describe glacial formations or the features made by glacial erosion date from the nineteenth century when Swiss and German scientists began serious geologic study of the Alps. An arête is formed when two or more cirques eat into a ridge from different directions. If these cirques, which expand by ice wedging as well as by glacial flow, cut through a ridge, a col is formed. After the glacier disappears, the craggy points rising above the col are called horns. The most spectacular example of a horn is the famous Swiss Matterhorn carved by three former glaciers.

In addition to shaping mountains, glaciers also alter the topography of valleys. Like streams and rivers that cut their own paths, glaciers tend to flow into existing depressions, but glaciers also blaze new trails, widening and gouging as they go. Because of their mass, glaciers reform valleys into U-shaped structures with broad floors, which can be easily distinguished from the unglaciated V-shaped valleys cut by mountain streams. In addition, since the ice mass is less fluid than running water, it tends to straighten out the twisting curves of a stream-cut valley.

When mountain glaciers reach the wide-open spaces, they form piedmont glaciers. Often a piedmont glacier is fed by more than one valley glacier.

The glacier may diminish by calving small icebergs in freshwater lakes or perhaps by retreating with a change in climate and leaving telltale terminal and lateral moraines to mark its advance. Occasionally, some of the ice may escape from the mountains to become outlet glaciers, which approach the sea, as has occurred along the coasts of Greenland and Antarctica.

*The Matterhorn (left) is a remnant of bedrock carved into a point by three passing glaciers. This area is studied for alpine glaciology. The water in the foreground is a cirque lake. The **glacial tongue** (right) formed in a mountain valley on Antarctica.*

Ages of Ice

Much of the modern history of the world has been heavily influenced by glaciers that marched across the face of the earth. As these ice sheets advanced, they brought with them colder, damper climates, so that many areas of what we now know as the temperate zone were almost subarctic in their weather. Cold-climate animals like the woolly mammoth and woolly rhinoceros roamed northern continents, areas that today would be much too warm for them.

The last great ice age began about a million years ago, marking the start of the epoch geologists call the Pleistocene. The advance of both continental and mountain glaciers continued until about 20,000 years ago, when they started a retreat that is still going on today. Only 6000 years ago, the continental glacier still covered large parts of Canada and Scandinavia. The history of glaciations is, naturally, directly connected with the large variations in the level of the oceans—20,000 years ago, the sea level was about 560 feet lower than today.

The disappearance of the great ice sheet was the beginning of the end for many of the animals that had successfully adapted to a glacial environment. In addition to the woolly mammoth and woolly rhinoceros, the mastodon, dire wolf, and saber-toothed tiger became extinct. But if the end of the Great Ice Age spelled doom for some animals, it allowed the emergence of others, especially man. *Homo sapiens,* or at least his anthropoid ancestors and relatives, had been living in areas that had not been covered with ice. After the glaciers began to retreat, however, man came out of the Stone Age and began using tools made of metal.

The Great Ice Age is usually divided into four periods of glaciation and three interglacial periods during which the ice retreated. In North America, the oldest is the Nebraskan glaciation, dating from the start of the Pleistocene. This was followed by the Aftonian interglacial, Kansan glaciation, Yarmouth interglacial, Illinoian glaciation, Sangamon interglacial, and the last advance, the Wisconsin glaciation. In Europe the seven sequences begin with the Günz, which coincides with the Nebraska, and is followed by the Günz-Mindel interglacial, Mindel, Mindel-Riss, Riss, Riss-Würm and Würm.

There were also much earlier periods of glaciation in the history of the earth, but their advances and extent are much harder to determine because of the subsequent modification of the earth's topography.

These icebergs (left) were calved by **Antarctica's** **ice cover,** *similar to the glaciers that advanced from the North Pole during the ice ages.*

Continental glaciation drastically altered the environment, allowing unique animal forms to flourish, such as the **woolly rhinoceros** *(below) that roamed areas which today are in the temperate zone.*

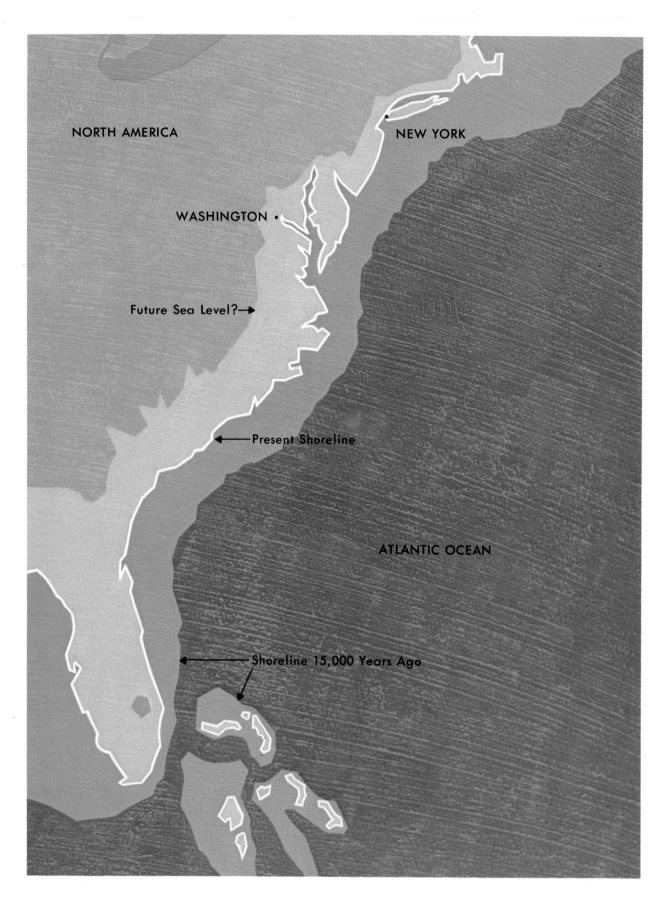

NORTH AMERICA

NEW YORK

WASHINGTON •

Future Sea Level? →

← Present Shoreline

ATLANTIC OCEAN

← Shoreline 15,000 Years Ago

Coming of an Ice Age

Among the effects of large-scale glaciation is a change in sea level, for water must evaporate from the oceans, cool, and crystallize to form snow. The snow then falls to earth and begins to become compacted into firn before becoming ice. The vast amount of water needed to accomplish this during the last glacial epoch lowered the level of the seas by about 600 feet below their present level, or to the ends of the continental shelves. Conversely, if all the water now trapped in glaciers and snow were to melt, the level of the oceans would rise by about 200 feet all around the world.

This lowered sea level allowed land bridges to be exposed not only in glaciated areas, where they might have been iced over, but also throughout the world. Lands that are now islands were connected to continents or other islands. The most notable examples of this are the Indonesian Islands and the land bridge that connected Siberia and Alaska. The Bering Strait was dry as recently as 12,000 years ago, and was used by both man and beast to migrate from Asia to America.

Since glaciers form in cool, damp climates, it is logical to assume that areas away from the glaciers were very much drier than they are now. The only alternative would be an increase in the total moisture content of the earth and atmosphere and there is nothing to indicate that this has occurred.

The shoreline moves with the tides, and it moves with the dramatic events of the earth and glaciers. We show here the present sea level for the east coast of the United States, and two other possible shorelines. During the glacial advances of 15,000 years ago, the shoreline of this area was at approximately the edge of the continental shelf, some 600 feet lower than it is at present. If all the ice on earth were to melt, the sea level would be approximately 200 feet higher than it is now, flooding New York and other low-lying coastal cities.

The earth has experienced many ice ages, and, in fact, it may be that at no time in the history of the world has there been a time when no ice was present. There were major glaciations in the late Precambrian period, 580 million years ago, and another in the Permian period, 225 million years ago. Since there have been several ice ages in the past, it would seem likely that there will be more in the future. The problem is that we know very little about the mechanics of new glaciers and glacial epochs, how they get started or why they end; so there is now no way of knowing if or when one is starting.

Some geologists feel that even now man may be artificially inducing an ice age by producing so much carbon dioxide through burning fossil fuels and at the same time reducing the total amount of land vegetation that utilizes carbon dioxide in photosynthesis. This extra carbon dioxide would hang above the earth, allowing ultraviolet radiation from the sun to penetrate and preventing infrared waves from radiating this energy back to space. This "greenhouse effect" may cause temperatures to rise to the point where existing glaciers would melt. This, in turn, would raise the level of the oceans. As more of the earth's surface became covered with water, more of the sun's rays would be reflected back into outer space, since water reflects solar radiation much better than land does. The result would then be a gradual lowering of temperatures until an environment conducive to glacial formation would be produced. But it has not been proved that industrialization actually raises the carbon dioxide content of the atmosphere to any substantial degree. It is often assumed that the oceans play a buffer role, and that they can easily absorb and chemically combine far more carbon dioxide than man can ever produce, unless, of course, something interferes with the sea's buffering capabilities.

Making of an Ice Age

Glaciation of the proportion needed for an ice age requires a powerful mechanism on a global scale. There have been many suggestions about the nature of this mechanism, but little proof. One suggestion is that the earth, as it rotates on its axis, wobbles slightly like a spinning top. This wobble, as much as five per cent from its normal precession, has been confirmed to the satisfaction of most skeptics. Some theorists say that as the earth wobbles back and forth in geologic time, the distance between the earth and the sun increases and then decreases. Thus the amount of solar radiation reaching the earth's surface varies. This, it is said, would account for the heating and cooling needed to move glaciers back and forth.

Cosmic dust has also been mentioned as a catalyst for global glaciation. Large clouds of cosmic dust do exist in outer space, and if the earth should pass through one, the theory goes, the sun's rays would be sufficiently screened to allow temperatures to drop and glaciation to begin. There is no way to disprove this hypothesis, but there is also no way to prove it, unless the earth approaches a cloud of cosmic dust in the near future.

There are some geologic phenomena that we know have occurred, but their effects on glaciation remain pure speculation. These include paleomagnetism, wandering magnetic poles, and reversal of the north and south magnetic poles. The earth's magnetic poles, and the whole magnetic field for that matter, are at the same time mysterious yet well known. Magnetism is known to induce the Van Allen Belt, which surrounds the earth and protects it from electrically charged particles from outer space. Yet this field—the north-south orientation around the earth—has alternated once or twice about every million years. The forces behind these pole

reversals and whether the switches are instantaneous or prolonged are a mystery.

There is no lack of evidence that the magnetic poles have flip-flopped, for data gathered from the ocean basins clearly shows this. As igneous rocks are pushed up from inside the earth, they cool and while they solidify, they are magnetized in the direction of the earth's magnetic field at that time. As more molten rock is pushed up in the midocean ridges, and as the earth's magnetic field has switched, the orientation is completely reversed. The process has continued and the seabed rocks show a zebra-stripe pattern of alternating magnetic orientations, first in one direction, then in the other, providing the evidence for pole reversals in the past.

Just as the poles have reversed, so have they wandered. The magnetic poles do not coincide with the geographic north and south poles, and they certainly are not in the position that they have always been in. Conti-

Ice can dramatically transform a landscape, as has this near iceberg located on the shore of Hudson Bay not far from Cape Churchill, Manitoba.

nental volcanoes spew molten rocks, which harden and adopt the magnetic orientation of the earth's field at that time. Subsequent eruptions, perhaps a few thousand years later, and more igneous rocks set, adopting a slightly different orientation—not enough to show a complete reversal of the poles, but enough to show that the poles had moved between eruptions. Even in historic times, navigators and astronomers have noted differences in their compass readings when compared with old charts and records.

The forces behind this erratic behavior of the earth's magnetism are still a mystery, but since they are on such a grand scale, many geologists believe that they must have global effects, and ice-age glaciation may very well be one of these effects.

49

A Part of the Drift

Much of reconstructing history is a game—putting together pieces of evidence and filling in the gaps with educated guesses. One of the bigger puzzles in the history of the world is the position of the continents. Why are these landmasses so concentrated in the Northern Hemisphere? It is interesting to note that more than 2000 years ago, scientists of ancient Greece knew that the earth was spherical and rotated around the sun, and they had computed its diameter fairly accurately. They somehow had knowledge that most continents were concentrated in the Northern Hemisphere and had made the amazing assumption that, in order to keep the globe in balance, there had to be a large continent at the South Pole.

Through geologic dating, it has been learned that the ocean basins are much younger than the continents. There have been fossils recovered on land that are older than 600 million years, but the oldest findings in ocean beds date only from the Mesozoic era and thus are no older than 200 million years. And why are the continents so much older than the seas? Very similar fossils have been recovered from different continents that had been separated for at least 100 million years, more than enough time for divergent development, even if the fossils' ancestors were of the same stock. But the question is: How could such distribution occur?

One of the proposed solutions to these puzzles has been that the continents have moved about the surface of the globe. The landmasses are made up of rocks that are much lighter than those that underlie the oceans. The theory of continental drift (See Volume

Fossil tree trunks and ancient coal beds prove that Antarctica was once a land with a temperate climate capable of supporting much plant life.

50

XI, *Provinces of the Sea*) holds that all the land areas were once grouped together in one, or perhaps two, supercontinents that began to break up and slide in various directions to their present positions. This theory was first proposed in the middle of the nineteenth century but didn't gain much popularity until our own time. One of the early "proofs" was the fit between certain shorelines. South America and Africa looked as if they could be placed together like pieces of a jigsaw puzzle. With a little fudging here and there—to allow for erosion, change of sea level, and local land uplifting in a few places—any number of reconstructions could be made.

In the first third of this century, however, scientists began finding stronger evidence, especially in the Southern Hemisphere, and Antarctica was a key component. The southernmost continent has not always been ice-covered, as coal deposits discovered there indicate, since coal is formed from decaying organic materials which require a much milder climate in which to flourish. Two of the plant genera found in these coal beds are the extinct *Glossopteris* and *Gangamopteris*. Abundant fossil evidence of these plants—as well as coal and glacial tillite—has been found in the same sequences in beds in places as widely separated as Argentina, Brazil, India, Australia, Tasmania, South Africa, the Falkland Islands, and Antarctica.

In addition, fossil remains of an extinct amphibian relative of the frog and salamander have been found which may strengthen the evidence that the above-mentioned southern lands were once linked. A bone of a labyrinthodont more than 200 millions years old was found in Antarctica, indicating that at least semitropical animal life was abundant there during this period. The other continents and islands, of course, have living relatives of the labyrinthodont.

The theory of continental drift generally holds that most of the southern lands were joined with Antarctica. Then as they started to drift apart, Antarctica stayed at the pole, rotating slowly while the other continents made much greater strides.

*A small piece of **jawbone of a labyrinthodont,** an extinct group of amphibians similar to the remains of the species below, was found in Antarctica.*

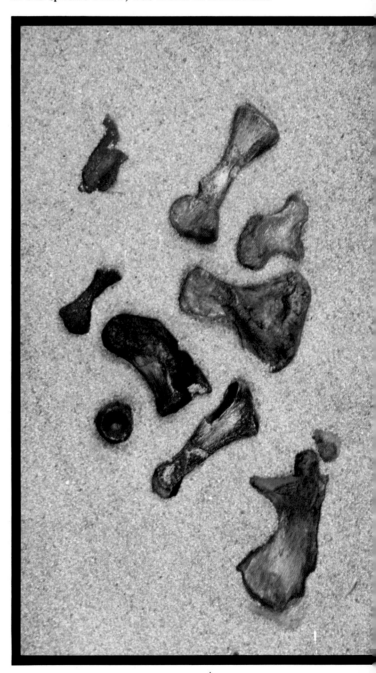

Chapter IV. Life in the Balance

Life in the arctic is the story of predators and prey, hunters and hunted. The polar bear ranges about the pack ice in search of seals, but will also battle toothed whales and even walruses, although in these latter engagements the bear could well be the loser. When traveling inland, the bear will even set upon the tiny lemmings. Wolves, foxes, and toothed whales are all carnivorous and prey upon other beasts. The caribou, musk ox, and snowy hare are their victims, as are the seals and lemmings.

Yet life is not constantly an eat-or-be-eaten struggle. The limited number of species in the arctic indicates that the territory—however large it may be—is not big enough for vast numbers of creatures competing for the same small amount of food. Polar bears stay away from the inland areas populated by

> "Things get out of hand only when the lemmings overpopulate an area, as is their wont from time to time, and then commit mass suicide in the sea."

the caribou and musk ox. The large herbivores, in turn, developed separate habits, with the caribou migrating seasonally in search of food, and the musk ox developing special defenses against wolves and cold that allow it to live a more sedentary life.

Among the smaller animals, wolves and foxes divide the territory so as to reduce the competition; and the hares and lemmings, though both rodents, do not often impinge upon each other's existence. Things get out of hand only when the lemmings overpopulate an area, and then commit mass suicide.

But the edge of the land presents a different story. Life in the sea is abundant; vast numbers of seals and walruses survive offshore, while mammoth whales roam about in its depths. Ice floes may not be a frozen paradise, but life certainly is not as difficult and uncomfortable there as it is on the arctic land. Only the unnatural predator, man, has upset the balance through his centuries of sealing and whaling activities.

Among the unique aspects of the arctic environment, and there are many, is the way sound travels. The air of the cold, dry, and flat northlands transmits sounds for great distances. Noises, even as gentle as a man crunching through dry snow, can be heard afar, and at times seemingly amplified. The sound of a gun firing does not frighten caribou since the sharp bark sounds like the snap of a snow-laden bough cracking.

The marine mammals are especially noisy, both in the water and out. On land the bulls bellow at the cows in order to keep them in line and snort and grunt at male rivals. Many, of course, have vocal danger signals and roar when approached by an outsider. In the water several species of seals have been recorded emitting bleats and beeps that are used for echolocation and perhaps even for communication. Among the whales the beluga has been known as the sea canary since the days when whalers first detected its vast range of vocalizations, which include clicks, whistles, cackles, trills, and whinnies.

It may not have been easy, but the large animals of the northern ice and tundra have been able to find a place for themselves.

The ruler of the arctic lands of snow and ice is the **polar bear,** *one of the largest terrestrial carnivores. The woolly white beast is most often found on the pack ice where land meets the sea.*

will show no sign of melting, indicating that there was no great transfer of heat from inside the seal's body to the ice. In the Antarctic, where seals have no Eskimos and no polar bears to fear, they sleep deeply on the ice and are easily approached by men.

In the spring, when seals lie in the sunshine on the ice, their skin absorbs solar radiation on one side of the body, while the other side is still cool enough not to melt the ice. Two experimenters once took the temperature of a seal which stuck its back out of the icy water into air that had a temperature of 55° F. Within minutes the part of the seal's skin that was exposed had a temperature of 62°, while the part that was still in the water was only 36°. The seal accomplishes this by closing off the blood vessels near the surface of its skin from the main circulatory system. Occasionally, enough blood is let through to nourish the skin and to keep it from freezing.

Seal Heat

Seals spend most of their time in the chilly arctic water, and they have developed a remarkable internal heating system. Many animals, including the seal, have the ability to allow their extremities to cool in order to preserve heat in the central core of the body. Seals are able to heat or cool different parts of their skin as necessary.

Seals often come out of water—which at its coldest is a few degrees below freezing—to take short naps on the ice. These naps are usually less than a minute long; then the the seal wakes up and spends 10 or 15 seconds scanning the horizon for predators, mainly polar bears or Eskimos. After the survey, the seal takes another quick nap and repeats the procedure. At the first sign of danger, the seal slips back into the water. The spot on the ice where it had been resting, however,

This thermoregulating ability is exhibited only by the phocids, or true seals, which are covered with only one coat of hair. The phocids include the hooded, elephant, and northern true seals. Their hair keeps them warm when they are out of the water by trapping an insulating layer of air beneath it. But when the hair is wet or when they are in the water, the phocids depend on blubber.

The fur seals are kept warm by their two thick layers of fur. The outer coat is long and coarse, while the underlayer is short and soft. Fur seals also have a thick layer of blubber, but do not have the temperature-regulating skins of the phocids.

The waters of the arctic coasts are home for the ribbon seal (opposite, top), northern fur seal (below, left), Priboloff seal and, until they were driven from the arctic to take refuge on Guadalupe Island, the northern elephant seal (right and below right).

Forks of Ivory

The walrus is a true northern animal which was once found in great herds on islands and on ice floes in the north Pacific and north Atlantic as well as the Arctic oceans. No southern species is known, and the walrus family is so specialized that there is only one genus, *Odobenus,* and one species, *rosmarus.* The walrus has been described as intermediate between its close relatives—the sea lion and the true seal. But in reality, the walrus is a highly specialized offshoot of the sea lion family. Unlike its relatives, however, the walrus has almost no hair on its hide and must depend entirely on blubber for insulation.

One of its special adaptations to the arctic environment is teeth. Most walruses have 18

teeth, with the upper canines greatly enlarged to form their distinctive tusks. The walrus must live in areas where there is plenty of shallow water—that is, less than 200 feet deep—for it feeds almost exclusively on the forms of life developing on the bottom, including shellfish, especially clams and mussels. The modern range of the walrus is now for the most part limited to Alaska's Arctic and Bering coasts in the west and much further east along Hudson's Bay, Baffin Island, and the west coast of Greenland. So specialized have their feeding techniques become that the walrus cannot chase fish and squid like some seals, nor can it use its forepaws like hands to smash heavy shells with stones as the sea otter does.

The walrus, which can grow over 12 feet long and weigh up to 3000 pounds, has most of its bulk concentrated in the front half of its body. While it looks like a bean bag and moves clumsily on land or on ice, it becomes a powerful, efficient, and fast swimmer under water, where it even displays elegance. It was believed that the walrus feeds by raking the bottom with its tusks and using its mustache of whiskers called vibrissae as feelers. These whiskers certainly act as sensitive fingers, most useful in murky waters where visibility is nil. But modern theories and the study of stomach contents reject the use of tusks to plow the bottom and minimize the proportion of shellfish in the diet of the walrus. Starfish, urchins, crabs, and even alcyonarians are also part of its meal.

*The walrus's strong **herding instinct** (above) offers protection against the elements and most natural predators, but not the hunter with a gun.*

*There is some doubt as to whether or not the walrus **uses its tusks** (opposite) in feeding.*

Louis Prezlin of the Calypso *crew (left) **teaches a young walrus** that lost its mother how to swim.*

Walrus courtship (above) is not interrupted by the strife associated with harem protection. Eskimos have long **hunted walrus** (below), but the modern method is with a rifle from a motor boat.

Life and Death on an Ice Floe

Most pinnipeds exhibit similar behavior in breeding—ritualized fighting to defend a territory, for example. These battles are more bluff and gruff than bite and tear, but occasionally an encounter gets out of hand. Sexual maturity among the female walruses is reached between five and seven years of age, about a year earlier than in the males. As with all pinnipeds, the social order of the walrus is determined by the locality for copulation. But even though they mate primarily out of the water, walruses do not practice a harem system, although they are polygamus.

Mating usually takes place in May but may occur in April. The gestation period is about a year long, so it is not uncommon, especially for older cows, to skip a year between offspring. The slate-gray pup weighs about 100 pounds at birth and is nursed for a year to 18 months. Immature walruses stay with their mothers for as long as three years, and the females' rare displays of hostility are usually in defense of their young.

During migrations north, when the pack ice starts to break up in the spring, the walruses segregate themselves by sex while they travel on the floes. Older males are grouped together, and females and immature individuals of both sexes travel in another group. But not all walruses take to the floes each year, and there is some indication that a few head south, for walruses have been taken as far south as England and Japan.

Walruses live on the ice even more than polar bears do. They have surprisingly small stomachs for such large beasts and dive repeatedly after their favorite shellfish, although they will settle for sea snails, starfish, shrimp, and sea urchins. Some walruses are said to be hunters of large game. These may be individuals living in an area where shellfish are scarce, or they may be "rogues," walruses which were orphaned before they were able to sniffle for food on the sea floor. These rogues had to sustain themselves in any way they could, usually by eating fish and carrion. Eskimos tell many tales of walruses, usually rogues, killing seals for food.

Since both are large and inhabit the same territory, walruses and bears frequently come into conflict. If they meet in the water, the walrus has the advantage since it has free use of its tusks and the bear must use all four feet for swimming. Ashore, however, it is an even battle unless the bear uses a technique that many Eskimos claim to have seen. These stories, perhaps based on an old legend, have the polar bear sneaking up behind a sleeping walrus, picking up a chunk of ice and hurling it with its front paws at the walrus, smashing its skull.

The horn of the **narwhal** (above) is one of its front teeth. Rarely photographed, narwhals are combative (opposite, bottom) in their polar home.

Under the Pack

Of all the whales living in the Arctic Ocean, the most distinctive is the legendary narwhal. One of its front teeth has grown so prominent in the forehead of mature males that it has earned it the name "unicorn whale." Narwhals are only 12 to 15 feet long and the tooth, which usually protrudes on the left side, is more than three feet long and sometimes grows to two-thirds of the entire body length. What use is actually made of this tooth is uncertain, but suggestions range from an ice-breaking tool, to a seabed rake, to a sexual display symbol.

Natives hunt the narwhal for its spectacular ivory tooth and also for its blubber, which is relatively thick for an animal of its size. Hand-held spears and harpoons are the traditional weapons, and the hunters need plenty of line, for a stuck narwhal will plunge as deep as a thousand feet. It soon surfaces, however, and is easily hauled in.

Another distinctive whale found primarily in the Arctic is the ivory-colored beluga, or white whale. These 10-to-15-foot-long creatures lack a dorsal fin and instead have a small hump in the middle of the back. The beluga often migrates up freshwater rivers but is more often seen in coastal waters, where it can be trapped in a beach seine. The beluga is a good diver, as its diet of flounder, halibut, and other bottom-dwellers indicates. When they return to the surface, belugas make so much noise that they can be heard for some time before they break water.

That wide-ranging master of the seas, the orca, is also found in the arctic. The orca is the most powerful and most intelligent of all dolphins. Since the orca feeds on fish, dolphins, seal, and just about anything else it comes across, Eskimos on the ice display a certain wariness. If they suspect orcas are in the area,

they often stick a pair of spears or harpoons a foot or two into the water and make grunting, snorting, roaring sounds, hoping to convince the orca there are walruses around. Although orcas have taken walruses, it is a mammoth struggle; so the orcas would just as soon retreat as fight.

There are a number of other whales which roam the Arctic Ocean beneath the pack ice, including the scarce Greenland right whale, some of the rorquals, and the bottlenose whale. This latter, which ranges in size up to 30 feet, is found throughout the arctic searching for its favorite food, squid. Like the Greenland right whale, the bottlenose was heavily hunted in the last century.

*The **orca** (top) is the master of the sea, ranging far and wide and feeding where it wants to.*

*The **beluga or white whale** (above) is a noisy creature that provides tasty food for Eskimos.*

Great White Hunter

Polar bears are constantly on the move in search of food, resting occasionally in the snow and perhaps denning up during especially severe weather conditions. The only bear that makes any sort of elaborate den and remains in seclusion for any length of time is a pregnant she-bear.

The polar bear is really the great hunter of the north, for it ranges on land, ice, and sea seeking food. It will do battle against walruses, small whales, and, of course, its favorite, the seal. Bears are reluctant to enter water but once in it are so mobile that some scientists consider them aquatic mammals. On land the bear may range inland eating vegetation and allowing its claws to grow longer for another season on the ice. Although small foxes and hares prove to be elusive, lemmings—especially when they appear in great numbers—are taken by bears, as is an occasional salmon during spawning season. Birds are ordinarily safe, but their nests are not, for polar bears delight in smashing eggs and extracting their contents. The larger land beasts, the caribou and musk ox, seem safe enough, especially since they inhabit a different part of the territory. Even if their paths should cross, a solitary bear is rarely a match for a larger band or herd of beasts.

The bear is certainly omnivorous. But it is primarily a blubber eater, and seals are the tastiest, most numerous, and easiest-to-catch source of blubber a bear can find. The polar

*The **polar bear's** scientific name is Ursus maritimus, and this implies that the beast can often be found in the chilly waters of the arctic region.*

bear is well fitted for hunting aquatic animals from a base on the ice, since it has an exceptionally well-developed neck, shoulder, and forearm. Generally rangier and leaner than its more southerly relatives, the polar bear is capable of delivering a killing blow with its left paw and hauling out the dead seal in almost one motion.

The bear also has a well-developed sense of smell, which comes in handy during the spring when a female and young cubs are ravenously hungry. It is also the time of year, of course, when pregnant seals are excavating calving chambers in the snow above the ice. The cow begins digging into the snow with her flippers, pushing the unwanted material down the breathing hole until she has an opening four or five feet long by three or four

feet wide and only about two feet high. At the surface, there is very little sign to betray the location. Polar bears, however, are able to sniff out the seal chambers, unless the snow cover over the excavation is more than three feet deep.

After uncovering a newborn seal pup that is unattended by its mother, polar bears exhibit another extraordinary hunting trick. The bear takes the pup by its flippers and holds it in the water where it begins to thrash about, since seals cannot swim until they are at least two weeks old. The cow rushes to the aid of the pup, which she assumes slipped into the water, and the polar bear pulls the pup back out of the water and then hits the mother in the head and draws her out onto the ice through the breathing hole.

*Land, or at least ice, is where a polar bear prefers to be since this is where it can find **seals to eat** and have mobility to avoid the larger walrus.*

Hunting Parties

The polar bear almost always hunts alone, unless it happens to be a female accompanied by her cubs. But the bear rarely lacks for company—however unwanted—for the arctic fox and usualy a raven may tag along for the feast. If these hangers-on get too pesky, bears will often throw pieces of meat toward them to keep them quiet. An oddity of this throwing ability is that the bears always use their left paw. In this and a number of other ways, polar bears show a tendency toward "left-handedness."

The polar bear is so closely attended by the arctic fox that this little creature, no bigger than a common housecat, has been called the jackal of the north. The fox does nothing to aid the bear in its work and may actually rest while the bear takes up a vigil beside a seal's breathing hole. But once the kill has been made, the fox is alert, hoping that the bear isn't too hungry and will be satisfied with the blubber and leave the meaty parts. The fox is content with picking the skin and bones, though, and in difficult times even settles for eating undigested matter from the bear's fecal droppings.

The fox is a good hunter in its own right. It is able to kill a seal in her calving chamber by surprising her and striking directly at the fleshy area around her mouth. The seal carcass is too large for any one of the furry white foxes to consume, and soon the snow is filled with tunnels as more and more foxes join in the feast. Another prey of the fox is the lemming, which generally feeds on underground roots. When not feeding, the lemming finds the going much easier through the snow. A fox trots along, listening for the lemming's movements; once it has detected a lemming, it springs into the air and, with its paws and nose close together, dives into the snow. If its aim is true, and it usually is, the lemming is a goner.

Another predator of the north is the polar or arctic wolf, with its massive head and jaws. They can grow as long as six feet and weigh more than 150 pounds. Packs of wolves are rare and usually found where there is a large amount of carrion. The wolf prefers to hunt alone or in groups of two or three. Its main prey is caribou, and so the wolf is found on the grazing lands of the tundra and especially in the large islands of the Canadian Archipelago. When caribou is scarce, wolves live on a diet of carrion and lemmings.

Arctic foxes (above) *have a commensal relationship with polar bears. The fox partakes of the bear's kill much as a jackal finishes the lion's meal.*

Wolves of the northern regions (right) *are becoming very rare. Though fearome objects of legend, no human death can be attributed to the wolf.*

The Victims

One of the most sought-after prey of the arctic world is the little lemming, barely five inches long. The lemming breeds very quickly; its usual litter of eight is born after a gestation period of just 21 days—the shortest gestation among mammals. And the offspring are ready to breed just 20 days after birth. Within a year a pair of lemmings and the generations of their offspring would number 170,000,000. Such overpopulation obviously works against the lemmings because the more lemmings there are, the less food and shelter there is for all of them.

The arctic hare, despite all the traditional tales of rabbits' breeding habits, has no problems with overbreeding. Its major predator

is the fox, and one of the few times the hare is relatively safe is in the heavy snows of early spring, where its speed and elusiveness are particularly effective. The hare itself is a vegetarian, feeding on roots and shoots that it can extract with its tweezerlike incisor teeth. But in extreme cases, such as after waiting out a storm by crouching in a rocky crevice, a hare may be driven to eating meat from a fox trap or even a human storehouse.

The musk ox is an unlikely candidate for the victim category. These small but powerfully built beasts live in the northerly parts of Greenland and on the arctic islands of Can-ada and Alaska. They are grazers which thrive on the summer flora which blooms below the glaciers and between the bands of drift snow. Even when snow covers the ground, a powerful kick or two from the musk ox's sharp-pointed hoof and the cover is cleaved to reveal tender grasses. In par-ticularly hard times they are able to live off the internal fat reserve they acquire while feeding nearly 24 hours a day on the lush herbage of summer.

Man has victimized the musk ox much more than any natural enemy, and ironically the beast's defense is its weakness. When musk oxen sense danger, they "square up," with the older bulls on the outside of the group and the cows and calves inside. Their heads are lowered with their menacing horns providing the first line of defense. Such an awesome barrier is effective against most animals, but a man with a gun can pick off the oxen, be-cause the animals don't know how to flee.

Facing a danger as it does a storm, the **musk ox herd** *(below) forms a ring or semicircle, sharp horns out. A line of defense must do when there are not enough in the herd to form the circle.* **Arctic hares** *(Lapus groenlandicus) (opposite, bottom) and* **lemmings** *(Dicrostonyx groenlandicus) (opposite, top) share the same specific name and often the same fate—food of the bear, fox, and wolf.*

*Despite its impressive antlers and large body size, the **caribou is a herbivore,** wandering from range to range in search of food.*

Wanderer of the North

One of the most spectacular natural sights that could be witnessed in North America took place in the Canadian arctic when the caribou were migrating. As recently as the mid-1930s, people could see thousands upon thousands of caribou moving along with a gentle lope in a stream that appeared a half-mile wide and that could run on for over an hour. At one time there may have been as many as 2.5 million caribou, but with the coming of the white man and his rifle, the population was soon reduced to less than one-tenth that number. More recently, however, the number of caribou has been increas-

ing, although not approaching the vast population of the past.

Like most large migratory herbivores, caribou move about on a seasonal basis in search of food. The sedges, willows, birches, and grasses that provide food during the summer are not sufficient for the cold, hard winter, and the vast herds move further south to higher ground where lichens are exposed on rocks or can readily be found beneath the snow. They may travel as much as 300 miles between their summer and winter ranges, using the same trails year after year.

Wolves, waiting to single out and attack a lagging member of the herd, are an ever-present danger to the migrating caribou. But

especially in the spring, unexpected water is another hazard. There is a report that 500 caribou were drowned in a flooding river.

In times past the caribou meant the difference between life and death for many Eskimos. The meat was used for food, with the heart, liver, bone marrow, and tongue preferred. Head meat was a delicacy, as was a sort of sausage made from rendered back fat and stuffed into the animal's intestines. Quite often, the meaty steaks and chops were fed to the dogs. The hides provided warm clothing for the winter—perhaps trimmed with fur from the arctic hare—and sturdy tents in the summer when snowhouses would melt. The leather garments and tents were stitched together with the caribou's sinews. The thick skins of old bulls were prized for footwear.

*From the mountain valleys to the tundra, the **caribou migrate** each season. In earlier times massive herds of these beasts created "rivers of brown."*

Chapter V. Life on Southern Ice

As foreboding a continent as Antarctica is, it is not totally devoid of life. The vegetation is scarce, but there are various species of mosses and lichens which grow on the bits of exposed rock in the interior. And the tiny pink mite *Nanorchestes antarcticus* was found only 300 miles from the South Pole.

But no large animals live permanently on the continent, although both penguins and seals frequent the shores at various times of the year. The Adélie penguin breeds on Antarctica during the summer, and the larger emperor penguin mates and sees the winter through on the vast sheet of sea ice surrounding the landmass. The emperor has had to

---

"The Weddell seal, like the emperor penguin, chooses to stay south for the winter."

---

adapt to the severe wind and cold of the polar winter. Like other warm-blooded creatures living in the antarctic and arctic regions, the emperor acquires extra fat during the summer feeding season. And much like the arctic musk ox, the emperor has developed a huddling defense against the biting force of the galelike winds. The birds gather into a circular formation called a testudo, with the individuals on the outer edge of the circle presenting their backs to the wind. Inside the tightly packed birds have their heads tucked down and some of the individuals have as little as 20 percent of their body surface exposed. The harsher the storm, the larger the testudo. A group of Frenchmen headed by Jean Rivolier once observed all the birds—about 6000—form a testudo during severe blizzards at Point Géologie.

Penguins and seals inhabit the same general areas and display some common traits;

neither showed any fear of man when he first came on the scene since neither had known predators on the land. The Weddell seal, like the emperor penguin, chooses to stay south during the bitter winter, while the Adélie penguin and the crabeater seal both move north with the ice pack. Only two of the three branches of the pinniped family are found in the Southern Hemisphere—the eared, or fur, seals and the true, or hair, seals. Walruses are absent below the equator. Fur seals, more mobile on land, are able to reverse their rear flippers and simulate a gallop like other four-footed mammals, while the phocids are more aquatic and act on land somewhat like fish out of water, moving about like huge caterpillars.

The seals living on or about Antarctica itself are members of the hair group: the leopard, crabeater, Weddell, and Ross seals. The fur seals, thought to be extinct after man's slaughter of them in the nineteenth century, are making a slow recovery and are now found in small groups as far south as the Antarctic Circle. A close relative, the elephant seal, ranges toward the warmer climes.

The story of the seals would not be complete without mentioning the slaughter that extended from the eighteenth century until the early part of the twentieth. Seals, both northern and southern, were killed by the hundreds of thousands to provide both fine furs and oil for industry. Those species that were not suitable for either of these purposes were used for pet food. The end of the story of slaughter has not yet been written.

Penguins are cautious when it comes to entering the water, for there may be some predator like a leopard seal lurking about. Often the birds will mill around the water's edge until one of them is jostled into the water. If he isn't eaten, the others join him.

Flightless Birds

Penguins can be called the most primitive of birds because they have changed the least of any bird since branching off from some reptilian ancestor. The fact that they can't fly is curious, since it is almost certain that they evolved from flying stock.

In many respects, the penguin more closely resembles a dolphin or seal than a bird, especially in the water where it is much more graceful than on land. The bird is a swift swimmer, and its wings, which cannot be used for flight, might easily be mistaken for flippers. Antarctic explorer and zoologist Edward Wilson, in his journal of 1902, wrote upon first sighting two penguins: "They looked very large indeed for birds, much more like a small seal, the one lying down, but the other was on his hind legs." These were emperor penguins, the largest of the more than 15 different kinds of penguins that inhabit the earth. Emperors may stand as tall as four feet and weigh up to 90 pounds after a good summer of feeding on the plentiful antarctic krill, those shrimplike creatures which also nourish whales and seals. Some of the penguin's ancestors may have been taller and perhaps bigger than the modern emperor; for five-and-a-half-foot-long fossil skeletons have been found that date back 60 million years.

The penguins inhabit only the Southern Hemisphere, but their range is surprisingly wide for flightless creatures. The majority of the species are found in and around Antarctica and the nearby subantarctic islands, but their relatives have reached Chile and

Peru in South America, the Galápagos Islands, South Africa, Australia, and New Zealand. The cold ocean currents along the coastlines of these places provide the kind of environment the penguins can live in.

Of all different kinds of penguins, only the emperor and Adélie breed on Antarctica itself. Actually, the emperor breeds on the sea ice that forms during the harsh austral winter. The Adélie, the gentoo, the chinstrap, and the royal penguins all have their traditional breeding rookeries and colonies on the islands or the coastlines of the Antarctic, but spend most of their time in the sea.

The lot of penguins is not easy, for in addition to severe weather conditions, they have had to put up with man during the last centuries. In times past, penguins were used for fuel—because of their high body fat content—to heat the pots of boiling whale blubber. Penguins had no reason to fear man since

their only occasional enemies were the leopard and fur seals and the orca, and they were easily caught and clubbed into submission. And today the more northern species—especially those in South America—are threatened by the growing fishing industry, increased recreational activity, oil spills by tankers, and water polluted by other means.

*Territorial defense by **macaroni penguins** (opposite, top) includes nipping with their beaks.*

*A resident of Antarctica itself, the **Adélie penguin** (opposite, bottom) displays "ecstatic" posture.*

*Grouping is natural to **emperor penguins** as adults (top, right) or babies (bottom, right).*

*The **Magellan penguin** (below), inhabiting Patagonia and the Falklands, is not a true antarctic bird.*

Ice Mates

Pity the poor penguin during breeding time —he sometimes finds it hard to pick out a female penguin. As part of the mating ceremony, the penguin makes an offering of stones, dropping them from his beak at the feet of his intended. Quite often, though, the intended turns out to be another male; sometimes even a human being, and all the effort is for naught.

Somehow, though, penguins do get together and mate. In most species the females lay two eggs, but among the emperors there is only one egg laid. This is most likely because the emperor chooses to breed in winter on a bleak nesting site—a vast gray sheet of ice floating above the dark sea. Like its parents, the chick-to-be will never know land; its life will be spent on ice and in the water.

Once the egg is dropped, the mother positions it on her feet and covers it with folds of skin hanging from her breast. Within three hours after the egg has been laid, the egg is transferred from the mother's feet to the father's, where the incubation continues. The hen, which has been without food for two months, waddles off for a while to feed in the sea—which may be some miles away by this time, depending on the growth of sea ice. The chicks are hatched by their fathers in the dead of winter when the winds are al-

most constantly gale force and temperatures dip to −70° and −80° F. The watchful parents provide for the young during this harsh period, and by the time the next winter rolls around, the chicks are big enough to take care of themselves. After hatching, the rookery is a noisy place; in addition to the thousands of squawking adults, there is the squalling of hungry youngsters waiting to feed on the regurgitated krill and fish their parents furnish.

The king penguin, slightly smaller but otherwise almost indistinguishable from the emperor, also lays but one egg, but has a more temperate breeding site in the mud flats of the subantarctic islands.

The Adélies, the gentoos, and the chinstraps are the only other penguins to breed on Antarctica. These birds are much smaller than the emperors, usually under two feet tall and weighing less than 20 pounds. They migrate to the nesting site in early spring, build a nest of stone, select a mate, and lay an egg sometime during November or December. The whole process takes two to three weeks. The male then takes over the incubating while the female feeds. After she returns, they alternate the chore and in about five weeks the chick hatches. The youngster then has most of the summer to grow and develop the strength to begin his pelagic life and migrate north before the ice begins to form.

*Guano rings the nest of incubating **Adélie penguins** (above). The white color indicates that the birds have been feeding mainly on fish.*

Adult emperor penguins *(left) stand guard over their gray-colored chicks.*

75

At Home in Water

The penguins are not particularly defensive birds, and even the Adélie tolerates the presence of man after a brief get-acquainted period. During the breeding season, skuas may snatch an egg or hatchling that is left alone too long, but otherwise the penguins have little to fear on land or from the air. It is on the edge of the ice and in the water that the only danger lies—the leopard and fur seals. The penguin is much more the water bird than the land creature, for travel on land is awkward and difficult. Penguins often resort to "tobogganing" down a snowy slope to the sea, sliding on their bellies like playful children. But there's no way to make the trip back to the nest easy. When they reach the water's edge, they take a close look before plunging, since a hungry seal may be lurking about hidden in the murky shore

waters. The entry is the only dangerous moment in the penguin's life, when it could be caught by surprise; the seal can barely outswim a penguin, and in open water the penguin easily outmaneuvers the seal.

From *Calypso's* helicopter we have repeatedly observed groups of penguins "porpoising" in a given direction, and passing close to either leopard seals or orcas, without changing their course and without arousing the interest of the powerful mammals. It is in the water that the penguin is most at home. The bird is a born swimmer, propelling itself through the water with its wings —some call them flippers—and "flying" undersea at cruising speeds of six to seven knots, with bursts of over 20 miles an hour.

As graceful in water as they are clumsy on land, penguins use only their feet to steer. They have the ability to leap out of the water, sometimes as high as ten feet; in exiting from the ocean the birds rocket toward shore and pop up onto a rock or ice floe and land feetfirst. Groups of penguins are frequently met at sea, hundreds of miles away from any shore, voyaging at a very constant pace and swimming very much like dolphins, clearing the surface every time they take a breath and opening their mouths wide for a fraction of a second. This "porpoising" type of progression is only used when penguins "make way." When they are hunting, their group scatters temporarily and they dive individually, often to great depths, to find the shrimp, little fish, and larvae they feed upon. The emperor penguin has been recorded diving to 800 feet and staying 15 minutes underwater.

Though they appear dignified while standing still on land or ice, **penguins** *(left) look comic and clumsy when they begin to waddle toward the sea. Once in the water, however, they display speed and grace to rival a sea mammal's, especially when leaping out of the water (below) while "porpoising."*

*Because they had no terrestrial enemies, seals of the antarctic show **no fear of man** and sometimes engage in playful activity with a human.*

Mammals of the Antarctic

Only about ten species of seals and sea lions survived the slaughter of the nineteenth century sealers who butchered true seals for their fat and fur seals for their pelts.

The rarest of the true, or earless, seals is the Ross seal, which is a rotund creature that lives on the pack ice. Its flippers are large and well under the body, while its bulging eyes are a distinctive feature. The Ross seal, named for James Clark Ross, emits not only the usual grunts of many pinnipeds, but also some trilling "coos" that have earned it the name of the singing seal.

The most numerous of the southern true seals are the misnamed crabeaters, which use their interlocking teeth to strain small crustaceans from the water in the same way whales do. Crabeater pups are born with a full set of teeth and nurse for only a day or two before taking off on their own, which may be dangerous for them. A large number of mummified crabeater skeletons, some of them 2000 years old, were found well inland on Antarctica, and some observers believe that these were young seals that wandered off and didn't get back to the sea before the ice came. But glaciologist Troy Pewe, who excavated many of the skeletons, believes that these were old or sick individuals that crawled inland to die.

The solitary leopard seal is widely distributed in the Southern Hemisphere, reaching as far as Australia and New Zealand, although more common in and around the subantarctic islands. It was believed until recently that its favorite food was young Adélie penguins, but it has now been established that penguins represent a very small addition to its diet; the leopard seal, like most other antarctic animals, feeds primarily on shrimp. The sea leopards have large, sharp teeth, but these animals do not live up to their reputation. When on land, Adélies amble past leopard seals without showing any apparent fear.

The Weddell seal, which lives in the inshore waters of the continent and surrounding islands, is one of the best studied of the southern true seals, because of its accessibility to scientists, because of the small number of individuals, and because it does not migrate. This seal stays in the higher latitudes for the whole year, spending most of the winter in the water, which is much warmer than the air. The Weddell has the ability to stay under water for up to an hour and can reach depths as great as 2000 feet while diving.

The elephant seal is also a true, or phocid, seal. It has sprinkled the South Shetland Islands and the Palmer Peninsula with small, very noisy colonies and is part of the antarctic environment. In the arctic it has been exterminated, and the only remaining colony in the Northern Hemisphere is now living in a subtropical climate at Guadalupe Island, Mexico, in the Pacific.

*The **dos pelos seal** has a double thickness of skin to keep itself warm. Fewer than 100 of them can be found today in their home on Tierra del Fuego.*

*The **Weddell seal** is one of four species of hair seals found in antarctic waters. The others are the leopard, crabeater, and very rare Ross seals.*

Breeding Battlers

Both southern and northern elephant seals belong to the same genus, *Mirounga,* but are separate species. The southern part of the family ranges among the subantarctic islands, especially Macquarie and the Falkland Islands, as well as all along the 1000-mile east coast of Patagonia. The northern elephant seal is not a polar animal; it breeds on Pacific islands off the coasts of California and Mexico and ranges only as far north as Prince of Wales Island, Alaska. In fact *Mirounga* probably has a great tolerance for various climatic conditions.

The southern elephant seal is the largest pinniped, ranging up to 20 feet in length and 8000 pounds in weight. Its most distinctive feature is its large proboscis, which gives it its name. The seal uses this nose as a resonating chamber to produce a variety of sounds, especially during the breeding season before the cows arrive and while the bulls are still establishing their territorial rights. The most frequent sound is a low-pitched grunt that sounds somewhat like an amplified bull-

frog's croak. Occasionally the bull ends this vocalization with a threatening sound, which is a vibrating noise that has been likened to a Bronx cheer or raspberry. The bulls also make a third sound—a trumpeting bellow —which is similar to the call of terrestrial elephants, hence their common name.

This constant cacophony is part and parcel of the territorial battles that precede the establishment of harems and mating. The fights are largely pushing and shoving, chest-to-chest encounters with much neck nipping, eye gouging, skin pulling, and nose biting. Sometimes large strips of blubber are peeled from an opponent; this causes pain but is rarely fatal. Older bulls are pockmarked with the black scars of ancient battles and the red splotches of more recent skirmishes.

These breeding bulls stake out their territory in midwinter and wait for the cows and mating season to come with spring. After establishing territories, defending them, servicing the harem, and protecting the cows, the bulls may then finally return to the water to feed themselves. It may have been as long as three months between meals.

Such a long period without food requires a vast amount of blubber, which is what led to the exploitation of the elephant seals and very nearly resulted in their extinction in the last century. The blubber from a mature male could yield up to two barrels, or approximately 700 pounds, of oil.

*Like most seals of the world, the **elephant seals** (opposite) know no terrestrial enemies and are not frightened by the approach of a man. The males (top right), with their elongated noses acting as resonating chambers, do most of their battling while establishing territories prior to the mating season, when the females (right, center) come ashore and gather in harems. The cows usually limit their vocalizations (bottom right) to issuing warning to the pups not to stray too far away.*

Losing Skin Game

The fur seals are much more adapted to a landed way of life than are the true seals. Of those that live in the Southern Hemisphere, none is really native to the antarctic continent itself. The southern fur seal ranges from southern Brazil and the Falkland Islands to Peru and the Galápagos. The New Zealand fur seal, as its name implies, is found on New Zealand and various subantarctic islands including Macquarie Island. The Kerguelen fur seal calls islands both above and below the Antarctic Convergence home, including the South Shetland Islands off Antarctica and the islands of Saint Paul and Amsterdam in the Indian Ocean. Other fur seals below the equator do not range even this far south, such as the Cape of Good Hope fur seals and their relatives, the South American sea lion, found along the continent and the Falkland Islands, and Hooker's sea lion, which frequents Macquarie Island and the Auckland Islands.

These pinnipeds have a very eclectic diet. Some live primarily on the abundant krill found in antarctic water, but they also catch octopus and squid, fish, crustaceans (mainly Galathea and king crabs), and climb rocks to enrich their dinner menu with penguins or cormorants. They breed on land and form harems during the mating season. The old bulls, called beachmasters, zealously guard their cows against intruding bachelor bulls. Their mobility on land may spring from the fact that all species have five movable toes of about equal length on their back flippers. They are also accomplished swimmers—not by using their hind flippers in true seal style, but by "flying" with their long front flippers, exactly as penguins do.

All the southern fur seals have the two distinctive coats of hair—the short, soft undercoat and the longer, coarser outer layer. This

double-coating of fur may cause overheating in exceptionally warm weather or after physical exertion, especially since they are not equipped with as elaborate thermoregulating systems as the phocid seals. Their only ways of expelling heat rapidly are by panting and by sweating through the flippers. In the warmer latitudes these seals are often seen on their bellies, flapping their flippers in an effort to facilitate heat loss and cool off.

The southern fur seals have all been endangered, and some still are, mostly because they have never recovered from the greedy, irresponsible slaughter of the nineteenth century. Sealers by the score came ashore and battered the unsuspecting seals, which had known no land enemy in the past. The clubbed seals, some still alive, were stripped of their skins. Pelts of young and old, male and female, were taken indiscriminately. If a cantankerous old bull attempted to defend his harem, a well-placed poke in his eye quelled the belligerence and allowed the slaughter to continue. Today fur seals are protected to a fairly good extent in Argentina, Chile, and Galápagos. They are under strict control and thriving in Uruguay and South Africa. Their return in Antarctica is threatened by the decision of several nations to resume sealing in the coming years.

The most dramatic case is that of the "Dos Pelos" variety, a stout fur seal with a specially fine, thick fur which was hunted even more than the others, and thought to be exterminated. We have spotted the last small colony of these seals, no more than a hundred individuals, on a remote island that would be very difficult to protect.

*The **slaughter of fur seals** (above) was only recently checked. Pups were clubbed to death, their parents shot. The killing tapered off only when the survivors were too few to make the hunt profitable. There are still some fur seals in the arctic region (opposite), but the south is almost devoid of them. The **dos pelos seal** of Tierra del Fuego (below) and a few other southern fur seals will make a slow comeback, if they can be fully protected.*

Chapter VI. In the Air Above the Ice and Water

Climatic zones are much more difficult to define on the ocean than on land. Generalized descriptions and terminology are not as easy to assign in the antarctic as in the arctic, which has far more vegetation, more animals, and a centuries-long history of human settlements and exploration. In the antarctic there is very little vegetation, much water, great inaccessibility, no native humans, and few observable animals. Those animals that exist sometimes range over so broad an area of the globe that they are useless as indicators of distinctive regions.

But as indistinct as their boundaries may be, there are definite environmental zones in the antarctic, including the harsh interior of the continent at the South Pole, the naked edges of the landmass, the islands offshore and

> **"Land-based predators were unknown until man introduced his rats and cats and dogs."**

approaching the Antarctic Convergence, and the coastal areas and isolated islets north of the convergence. Sometimes the weather conditions help define a climatic region; and sometimes a type of organism or its adaptations differentiates a zone of life.

Nature has little regard for man-made boundaries and birds are among the most cynical of creatures in this respect. The seabirds roam freely through the Southern Hemisphere, breeding here, feeding there, and visiting in the unlikeliest of places. Despite their wanderlust, abetted by long-distance flying capability, there is no bird which breeds at sea in floating nests, as the ancients once believed. All avian creatures find their way to some solid surface—usually land, but

in some cases ice—in order to court, mate, breed, and rear their young. Antarctica itself, with perhaps less than five percent of its surface permanently free of ice, attracts some species of mating seabirds, but for the most part they choose barren coasts and lonely islands farther north.

Birds of the order Procellariiforms—the tube-nosed birds—are frequently found in the area south of the Antarctic Convergence. This is not their only range; some great migrators, flying long distances, even cross the equator to avoid the harsh winter.

The severity of the antarctic environment is offset by the lack of predatory terrestrial mammals. Some of the larger and more aggressive birds, like the skua, harass the smaller species and raid their nests for eggs and hatchlings. But land-based predators were unknown until man introduced his rats and cats and dogs.

The southern seabirds—the petrels, albatrosses, and shearwaters—are well adapted for an oceanic way of life. Their extended wingspan allows for gliding great distances and obviates the need for frequent landings. And their webbed feet complement their tube-nosed bills in the water world. The effectiveness of these adaptations can be seen in the large populations of some of these seabirds; the single most abundant species in the world may well be Wilson's storm petrel, which breeds on subantarctic lands. As an order, the Procellariiforms are among the most numerous, if not the most numerous, wild birds in the world.

Birds of the antarctic region have always been considered free spirits—knowing few men and little danger and having an abundance of food in the sea.

The Wanderers

The most remarkable birds of the southern skies are the albatrosses, the gooneys of many a sailor's legend and superstition. With their extensive wingspan and their ability to fly for years without approaching land, the albatross is a familiar and magnificent sight in the seas north of Antarctica.

The two largest members of the group are the wandering albatross, *Diomedia exulans,* with a wingspan of up to 11 feet, and the royal albatross, *D. epomophora.* It is easy to assume that these birds grew to their huge size because of an absence of predators on the islands and coastal areas where they breed, but the albatross is not the largest of all birds, for among the flying birds, the swan and bustard have greater body bulk. But the lack of natural enemies allows the albatross the leisure of laying a single egg rather than the nestful that other species lay. The albatross egg and chick require such a long developmental period that a breeding pair can rear a chick only once every two years. For at the end of the first year, the recently hatched young are still nestbound, demanding food and attention—which disrupts any courtship and mating behavior the parents might engage in.

These large albatrosses usually live to 40 or 50 years of age, although some individuals are believed to be as old as 70. Despite the general lack of natural hazards in their environment, there are only about 40,000 wandering albatrosses in the world, and perhaps 20,000 of the royal. As adults, the two are so similar in terms of plumage that they are difficult to distinguish in flight.

The other members of the albatross group are much smaller than the wandering and royal. Just as sailors collectively lumped the two great albatrosses together under the label "gooney," the small albatrosses were all referred to as "mallemucks." These are dark-mantled birds and include the gray-head, black-brow, white-capped, yellow-nosed, and Buller's albatrosses, all of the genus *Diomedea.* The sooty albatross and the more southerly-ranging light-mantled sooty albatross both belong to the genus *Phoebetria.*

*Beautiful, ghostly gray, the **sooty albatross** (opposite) has a white half-ring to set off its eye. The **wandering albatross** (left) is the subject of many a sailor's legend and superstitions.*

Effortless Flights

All flying machines, whether natural or man-made, require lightweight construction and lifting surfaces that must be large at slow speeds and takeoff, and can be very small at high speed. Bats and pterodactyls used stretched membrane as construction material for wings and man experimented with canvas and metal. But feathers, with a generous surface area of very little weight, are an unmatched material for wing coverings.

Among birds, there are different wings for different purposes. They may be built for speed or soaring or quick take-off or gliding. The long, narrow shape of bird wings, airplane wings, and fish tails all have the attributes of high lift for thrust and of low drag for propulsion. The albatrosses, with their long, narrow wings, are gliders that use the force of the wind itself to aid them in remaining aloft. The dependency on the wind is so great that albatrosses find it difficult to become airborne in areas of calm; they are, accordingly, absent in the low-wind areas of the equatorial zone known as the doldrums. Only rarely have albatrosses been seen north of the equator or inside the Antarctic Circle; they choose to remain closer to the "roaring 40s," the latitudes of strong and constant winds. Most of the albatrosses stay between 20° and 60°, with only the sooty albatross regularly flying as far south as Antarctica.

Flying in the manner of an albatross makes use of aerodynamic principles so complex that to explain them fully is still impossible and would destroy the simple beauty of the bird's flight. The basic phenomenon used to the very best advantage by the albatross is the turbulence created by drag, or friction, where air and water meet. The stronger the wind, the rougher the surface of the sea, the more spirals, or gyres, are created. These spirals lift and lower the bird as it tilts one wing toward the earth like a banking airplane. The albatross banks one way, then the other, alternating in a gentle roll. There is only an occasional beat of the wings.

The albatross not only uses the wind as lifting power, but it also manages to remain master of its own course, either being carried all the way around the globe or, on the contrary, making way against the strong winds without a beat of its wings.

The gliding ability common to all albatrosses is somewhat modified in the sooty members

The **gray-headed albatross** (opposite) shows no trace of its wingspan because its wing folds in three parts instead of the normal two. The **giant petrel** (above) is often mistaken for an albatross.

of the family, for they are also capable of bouyant flight. With their smaller bodies, proportionally longer wings, and long, pointed tails, they are able to fly higher, change direction more quickly, execute more intricate maneuvers, and land more easily than the larger albatrosses, both the gooneys and the mallemucks.

Storm Birds

For centuries sailors have applied the name petrel to any number of birds they saw winging above the sea in the middle latitudes. Most of these creatures are small, dark seabirds with long wings and features generally adapted for living above and from the ocean. Scientists group them in the order of Procellariiforms. Although this name derives from the Latin word for storm, with which the sailors associated the birds, their most distinctive trait is their bills, where nostrils are located in a tube or in tubes.

There are many families of petrels, including the diving petrels, storm petrels, gadfly petrels, and fulmarine petrels. The diving petrels are the smallest group, in terms of numbers, represented in the subantarctic area by the South Georgia diving petrel and the Kerguelen diving petrel.

The gadfly petrels nest in underground burrows in zones where the soil does not freeze, but they frequent the antarctic region for feeding. The most southern ranging of these are the mottled petrels which can be found on the ice pack during the summer months, between migrations to the northern Pacific off the west coast of North America.

The fulmarine petrels nest in the ledges, rock piles, and crevices of exposed areas, but being fairly large, they are able to defend their eggs

Gulls are common sights at every seaport in the world and even visit the poles. This one was found over the Norwegian Sea at 68° north latitude.

To avoid preying fur seals, the **cormorant** nests well above sea level. Also known as shags, there are three species of cormorants in the antarctic region.

and chicks against the marauding skuas. The fulmarines themselves are scavengers and predators, with the two largest species about the size of mallemucks.

Two of the more familiar fulmarine petrels are the snow petrel and the antarctic petrel, both of which breed on the continent itself. The snow petrel is often found on the pack ice or the floating tabletops of icebergs, and returns to Antarctica in the summer to burrow among exposed rocks and nest. The antarctic petrel is not so solitary, and it breeds in colonies of several thousand.

One of the most abundant, and certainly the widest ranging, of the fulmarine petrels is the cape pigeon, also called the pintado petrel.

This bird, with its distinctive checkered mantle, is found from Antarctica to very near New Zealand and almost to South Africa. A real scavenger, it has recently flourished because of the carrion and refuse provided by the whaling ships that ply these waters.

Perhaps the best known of the petrels, to nonsailors at least, are the swallow-sized storm petrels, the smallest of the seabirds that breed in the antarctic region. These tiny creatures must nest in burrows in order to avoid the predating skuas. But despite their small size, the storm petrels are great travelers. Wilson's storm petrel, one of the most abundant species, flies as far north as Canada to escape the austral winter.

A rare color for its species, this **white giant petrel** *shows the complicated beak and nostril structure common to the Procellariiformes.*

A Bird by Any Other Name

Seamen were ready observers, quick to name the birds they saw, sometimes pinning two or three appellations on the same bird. In other instances, many different species were lumped together under one name for the sake of quick identification.

Common names, which differ from island to island and country to country, may be based on something as concrete as the color of a beak or may arise from a sailor's imagination. For an example, one man's petrel could be

Sometimes referred to as the antarctic eagle, the **predating skua** shown in the photographs on this page will attack young birds and unguarded eggs.

another man's shearwater. Many seabirds fly very close to the surface of the ocean, seemingly skimming on top of it. Hence the name shearwater is given to many birds. But other sailors, observing the same flying maneuver, were reminded of Saint Peter walking on the water, and named the bird after him—Little Saint Peter. Over the years this was corrupted into petrel. Albatross is supposedly derived from *alcatraz*, Spanish for pelican, while mallemuck was formed from words meaning foolish and gull.

After the sailors came the scientists who studied the birds and tried to bring some

order to the nomenclature. They couldn't agree among themselves in all cases, however, on names or on how to group the birds. As a result, there are birds like the prions which are petrels but whose broad-billed beaks elicited a Latin name meaning "whale birds," perhaps because the birds, like whales, feed on crustacea. However, as one observer wrote, "The prions are taxonomically an extremely complicated group of petrels and nearly impossible to identify at sea." The shearwaters present no great problems in the area of grouping, but their names are somewhat confusing. Some species are called chin-strap petrels and black petrels. In general, the smaller shearwaters breed in burrows dug into soft earth fairly far north of Antarctica toward South America and in the area of Australia and New Zealand. The southernmost is the sooty shearwater, which has been known to breed on Macquarie Island and feed almost to the shores of Antarctica during the summer months.

The skuas are distinctive scavengers, preying on isolated smaller birds and abandoned eggs, and especially on the Adélie and genoo penguin eggs and chicks. With its strong beak and curiously clawed—yet webbed—feet, the skua has been improperly dubbed the eagle of the antarctic. In fact, a skua is easily chased away from a colony by the smallest adult penguin and can make a living only by patiently waiting for the penguin parents to make a mistake. Skuas also deliberately attack humans approaching their nests, but they never strike—they attempt to intimidate and discourage the intruders, which is proof of conscientious parenthood and courage, not of viciousness.

Cormorants of the Southern Hemisphere (top) generally have white breasts and underparts. The double-crested cormorants of the Northern Hemisphere (left) are more uniformly black.

In the Wake of Man

Seabirds range far and wide to feed, because not all areas of the oceans are equally supplied with life, nor are the small crustaceans, squid, and fish that feed the birds equally abundant at all times of the year. But because of their flying range, the birds of the southern oceans do not have to compete strenuously for food. In fact, because of the scarcity of land and the harshness of the environment in such land, there is much more competition for nesting sites than there is for food.

But during the dead of winter, when ice covers much of the sea around Antarctica, food may be hard to find for the seabirds. For this reason, they wing after icebreakers as they roll about ploughing to and from the land, opening leads in the ice and affording access to the sea beneath.

Scavenger birds like the cape pigeon and other small fulmarine petrels also trail ships, hoping for scraps of food or sometimes just using the vessel as a temporary perch. The whaling ships, of course, attract many more followers than, say, *Calypso* or any other exploration ships. The giant petrels, cape pigeons, sooty albatrosses flock to a factory ship not only for the bits of whale blubber and waste material, but also to pick on the carcass of the whale itself as it floats beside the ship before processing.

There are endless stories told by sailors and travelers about their experiences with birds trailing their ships in subantarctic waters. Probably the most famous of these is related in verse by Samuel Taylor Coleridge in *The Rime of the Ancient Mariner.*

The Calypso *is on its way to Hannusse Bay, Antarctica, and since it is not equipped with an icebreaking hull, it must be guided carefully by helicopter as it negotiates a path through Gunnel Channel.*

Chapter VII. Riches of the Poles

No one would describe the polar regions as lands of plenty, but the waters of these areas are certainly seas of abundance. Perhaps the best indication of this is displayed by the existence of the many shrimplike creatures which provide food for the largest animals on earth—the whales—as well as for large fish, seals, penguins, and seabirds. Norwegian whaling men referred to the shrimp they found in southern waters as krill, and the name is applied to any of the small crustaceans which whales feed on. While some scientists prefer to use the term only in reference to those species in the antarctic waters, most say it refers only to a single species, *Euphausia superba*. Whatever the case, and we will use the term in its most general sense, the abundance of krill is one of the greatest riches in the sea.

This plentitude of krill has allowed many aquatic animals to develop specialized feeding styles that include filtering systems which screen the crustaceans from the water. The crabeater seals, discussed earlier, sift euphasiids from the sea. Breeding penguins

---

"Breeding penguins eat their fill of the tiny crustaceans, then disgorge the krill on land to provide food for the chicks."

---

fill themselves with krill and, like parent birds everywhere, bring the food back to their young. They then disgorge the krill and provide a meal for the chicks. It is the baleen whales which are the most specialized of the krill feeders. These huge beasts have enormous mouths that allow them to take in vast amounts of water. The seawater is then forced out through the sievelike baleen plates. The network of stiff plates, which is made of keratin, like man's hair and nails, and which takes the place of teeth, retains the crustaceans in the whale's mouth. They are then swallowed whole at a rate of several tons a day.

But the krill, and the creatures that feed on it, are the second-stage indication of the richness of the poles. The primary stage is composed of diatoms, the main food of the krill, exceedingly abundant in the antarctic. These and other species of phytoplankton abound in the polar waters because of the wealth of nutrients constantly brought up in surface waters from the deep "upwelling currents." It is estimated that, as a consequence of ruthless whaling, there is left today only 6 percent of the whales that used to feed in the antarctic, and that more than 150 million tons of krill are now available for other forms of life, which accounts for the recent proliferation of birds, penguins, squid, and fish, as well as for a population explosion of bottom creatures such as starfish, or ascidians. Russian and Japanese fleets are preparing to take their toll out of this bounty. The ice cover in the arctic prevents much planktonic growth there, but plankton and crustaceans are plentiful around the fringes.

Visually, the polar regions are storehouses of treasure, with the wind and water combining to form stark kinetic sculptures, illuminated by the exploding lights of the auroras. Uncapturable in print or on film, the beauty of the snowscape and the nightly light shows demonstrate the remarkable ability of nature to provide riches at its wont.

Caught in a water-trawl sampling about five feet below the surface, euphausiids spill out on the deck of a research vessel. The same haul could be obtained from the stomach of a baleen whale.

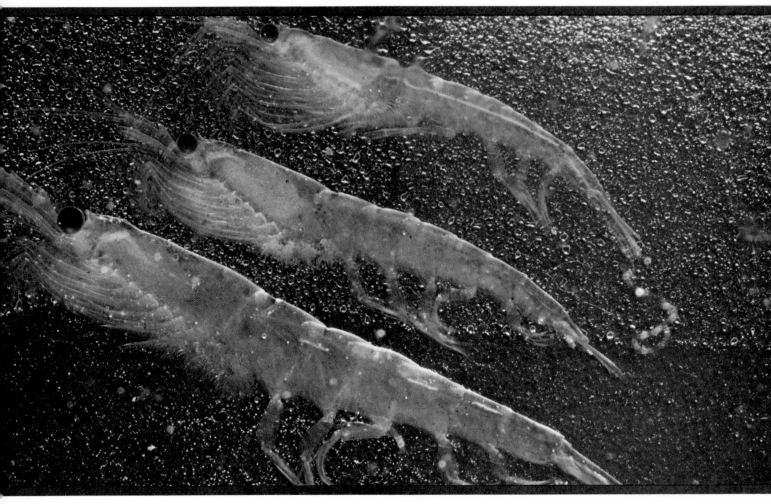

Euphausiids, or krill, (above) are plentiful in antarctic seas because of the abundance of diatoms
(below). These phytoplankton flourish because the
waters of the region are rich in mineral nutrients.

Soup of Life

We speak of bread as the staff of life because
it provides nourishment for so many people.
In the antarctic and, to a certain degree, in
the arctic, tons and tons of shrimp form the
foundation of life. Sometimes the sea in these
regions is so full of them that it resembles a
giant cauldron of pink soup.

Making up this rich broth are the euphausiids, tiny two- to two-and-one-half inch
crustaceans that feed on the abundant phytoplankton in polar waters. Although there

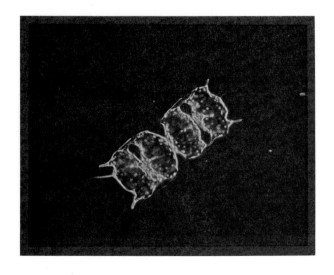

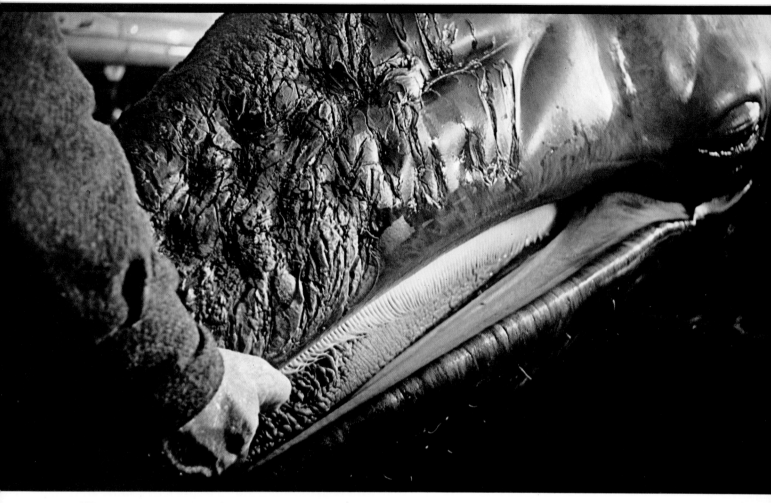

*The shrimplike euphausiids form the basic diet of the antarctic whales. The baleen in the mouth of a **baby right whale** shows the perfect device for straining sea water and extracting the krill from it.*

are many species, *Euphausia superba* is the most numerous and provides the bulk of, and in some cases the only, food for creatures of Antarctica itself, the subantarctic lands, and the waters south of the Antarctic convergence. These euphausiids, called krill, are commonly found only below the Antarctic Convergence, where the water of the upper 350 feet is cold, with large amounts of dissolved oxygen, and there is an exceptional supply of the silicious diatoms upon which they feed. Krill can live as long as 10 months in water at 32° F. and for shorter periods in temperatures below freezing. They concentrate in dense patches, some measuring only a few square feet, others up to a square mile in size. These patches are most common in a ring of cold water that surrounds the pack ice of Antarctica for a distance of some 600 miles. Where the westward flowing current of the Antarctic Ocean is deflected toward South Georgia by the Palmer Peninsula, krill becomes concentrated. There, in an area of about a million square miles, the sea is incredibly rich with food for larger fish, marine mammals, and birds.

Diatomic Living

To the uninitiated it is probably surprising to find out how abundant life is in polar waters. Somehow it is difficult to grasp that in the ocean the tropics are much less fertile than are the poles, where life teems.

Living requires food, and food anywhere—land or sea—begins with plants, for only vegetation has the ability to reproduce by drawing on chemical nutrients, carbon dioxide, and solar radiation. Animals, including man, receive most of their energy from the sun indirectly by feeding on plants or on animals that feed on plants. At sea, tiny floating plants, collectively called phytoplankton, are most abundant where the sunshine is most plentiful and where the nutrients are the richest.

Since much of the Arctic Ocean is always covered with ice, workings of the phytoplankton in colder waters is best known in the antarctic region, although the process does go on at both poles. In the south there is a deep-flowing body of water full of the phosphates, nitrates, and silicates that are

nutrients. Through the process of turbulence and upwelling, these nutrients are brought close to the surface, where phytoplankton explode in massive amounts during the six-month polar day, from September to April. During the time of 24-hour darkness, and when the sea ice thickens, growth of planktonic plants is inhibited.

Because of the peculiarities of deep-ocean currents—influenced by density and bottom topography—the nutrient-laden waters and resulting phytoplankton are more common in some areas than in others. The South Pa-

cific is a virtual desert, in terms of sea life, while coastal waters along the antarctic peninsula and the Weddell and Ross seas are fantastically rich, particularly in summer.

The most abundant phytoplankton is the diatom, which in some studies constitutes as much as 99 percent of the samples examined. But as with all plants, there are seasonal species; some are more plentiful in spring than in late summer, and others flourish geographically, being more numerous near the pack ice or further out to sea. During a recent exploration of antarctic waters, *Calypso* made continuous recordings of the chlorophyll content of sea water, which is indicative of the productivity of marine vegetal plankton, to be correlated with measurements made at the same time from satellites by chromatography. In the future, when space-borne instruments will be properly calibrated, the primary productivity of all oceans will be monitored by satellites without any need for ships.

Diatoms, which incorporate silica into their structure, are the most numerous. Also found in antarctic waters are silicoflagellates, as well as more than 30 other species of dinoflagellates, which are made of cellulose rather than silicon oxide.

Some of the phytoplankton is consumed directly by the larger creatures, but there are also numerous forms of zooplankton which feed on them, including the tintinnoidea. These ciliate protozoans number several thousand different species and are very good indicators of the antarctic waters, since virtually all of the species appear to be native to the southern-most oceans.

*The **Calypso crew** conducted a study of primary productivity in antarctic waters by measuring the phytoplankton and then sending the information to NASA's Ames Research Laboratory in San Francisco via a stationary communication satellite.*

Plants of the Poles

Snow, cold, sea ice, and low humidity—these conditions are common to both poles. And so we might expect that vegetation at the two ends of the earth would be similar, perhaps even closely related species. In truth, while similar plants appear in both polar regions, there do seem to be many distinct types of plants endemic to one region or the other. Any similarities would more likely result from adapting to the environment—such as the scouring action of wind-blown ice particles—than from close family relationship.

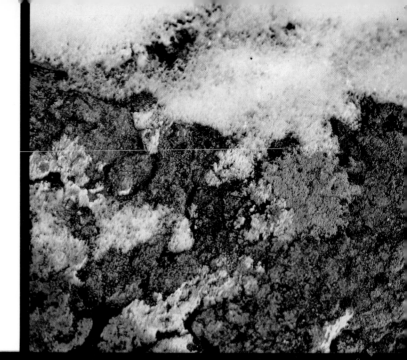

In the north the land areas are the ends of vast continents, so that vegetation could extend into the arctic from wide areas of temperate climate. Southern flora have few opportunities to spread from the small subantarctic islands and craggy coasts to rarely exposed lands.

It is not surprising, therefore, to find that the arctic and subarctic is much richer in plant life than the antarctic and at more extreme latitudes. On large arctic islands like Melville and Banks and even on the north of Greenland, there are white and flax saxifrages, dryas roses, stichwort, dwarf dandelions, cinquefoils, and poppies. The flowers are so numerous that bees appear in profusion. R. E. Peary reported seeing a bumblebee a half mile out into the Arctic Ocean.

On the tundra, of course, there is enough grass growing to sustain the grazing caribou and musk oxen, as well as the herbivorous

*Virtually nothing but algae can grow on the bare ice of the arctic and antarctic caps. But in the mountainous regions of Antarctica, **lichens** (opposite top and bottom) and a few other hardy species can gain a foothold. In the arctic, however, the islands surrounding the pole can support many types of plants, like the **figworts** of the Pribiloff Islands (above) or the riotous spring colors of Greenland (right).*

arctic hare and lemming. Trees are absent in the far north, but bushes of cranberries, blueberries, and crowberries are within range of polar bears summering inland.

Antarctica, except for its peninsular finger pointing toward South America, is devoid of flowering plants. Only mosses, liverworts, lichens, algae, and some fungi are able to subsist on the few bare patches of earth and rock. As in inhospitable places in the extreme northern arctic, the vegetation displays low growth and patchy distribution.

In Antarctica, lichens are easily the most widespread vegetation, and in some places they are the only growing plants. The mosses and liverworts living on the edge of the antarctic region are simliar to those elsewhere, while those bryophytes confined to the continent itself are more distinctive. The algae of the antarctic are both aquatic and terrestrial, with some types existing on the ice and snow, giving it a red or green or yellow tint. Many patches are large enough to be visible to the naked eye.

Of the flowering plants, only two kinds are found on Antarctica itself, and these types are restricted to the milder northwest coast of the peninsula. One is the pearlwort known as Antarctic pink and the other are two species of *Deschampsia* grass.

Visual Riches of the Poles

It has to be seen to be believed. This statement was never truer than when used in describing the natural beauty of the polar regions. Brutally cold weather on Antarctica fosters the thick ice shelf which spawns majestic icebergs, flat-topped and faced with sheer cliffs. These alabaster monsters drift in total serenity, their calm broken only by a cliff of ice collapsing in the sea, or by an occasional petrel punctuating a long day's

search for food with a brief stop. Ice floes in the arctic are often the setting for touching domestic scenes as mother and cub, or pup, move north for the summer.

The environment is harsh, but the beauty is there—in blue caves furnished with eerie interiors sculpted by meltwater or in the lonely snow dwarfs, peaked caps shaped by the wailing winds.

The most spectacular phenomenon, though, is not in the sea or on the land but in the air. Auroras—borealis in the north, australis in the south—provide dancing, swirling, undulating sheets of colored light. They cannot be described verbally. Even photographs are unsatisfactory, since the physics and chemistry of the forces involved render photography incapable of capturing the richness of color and fullness of beauty. Even an artist's palette lacks the flexibility of tone and value. And the occasional sighting of aurora borealis in Europe and populated areas of North America, once or twice a decade, is but a thin and distorted representation of the real polar thing. In other words, it must be seen to be believed.

But perhaps we can explain it. The aurora occurs, not because of the cold in the polar region, but because of an opening, a gap, in the earth's magnetic shield that allows cosmic radiation to penetrate. Solar energy in the form of highly charged ions and particles bombards the earth in the area of these holes and reacts with thin gases in the upper atmosphere, somewhat like neon tubes are activated by electricity to produce the light show. During the long summer days at the poles, the direct sunlight usually blots out the aurora. But during the six-month-long nights, the aurora can readily be seen shifting back and forth across the sky.

While the northern and southern lights are most likely identical, observers sometimes use different terms to describe them because of a difference in vantage point. In the arctic the aurora is usually viewed from land well south of the pole, so that the lights appear in the north over the ocean. In Antarctica, however, the display is directly overhead, and the change in perspective alters the beauty in the eye of the beholder.

*The spectacular color and lighting of the polar regions inspires photographers, yet at the same time eludes accurate capture on film, whether it is **aurora borealis** (left) or an **antarctic sunset** (below).*

Chapter VIII. Below the Ice

Life in the water world above the Arctic Circle and below the Antarctic Circle would seem to be much easier than life in the polar land or polar skies. And, judging by krill alone, this is at least partially true. There are certainly plenty of invertebrates, both pelagic and benthic, but there is also a relative paucity of fish.

In terms of inland waters, there are no freshwater fish in Antarctica, partially because there is so little freshwater. Many of the ponds and lakes that dot the continent are much saltier than the sea itself. What fresh-

---

Among the marine fishes there are probably no more than 100 different species and they show no bipolar distribution.

---

water there is exists as meltwater streams and lakes for only a short period of time each year. The life in these waters spend most of their time as eggs, then spring to life in the freshwater before laying eggs for the next season. Such animals include various protozoans, tardigrades, rotifers, and nematode worms. In the arctic, however, the conditions are less severe, and several fish are able to live in streams and subarctic lakes. These include pike, blackfish, and the arctic char, which spawns in many of the rivers flowing into the Arctic Ocean.

Among the marine fishes, there are probably no more than 100 different species, and they show no bipolar distribution—at least on the species level. At the South Pole the important fish are the antarctic cods, plunderfish, dragonfish, and icefish. In the arctic, the dominant groups are the sculpins, true cods, flatfish, and eelpouts. Although the anglerfish *Ceratias holboelli* has been found in the ant-

arctic region and off the coast of Greenland —and not in the tropical waters in between —bipolarity is more common at the genera and family level than at the species level.

One trait that appears common to most polar fish regardless of species is that they all lay their eggs on the ocean floor, apparently to avoid contact with the ice cover or with the low-salinity surface layer in the summer. The eggs are relatively large, with massive yolks for their size. When the larvae hatch, they are also fairly large and they can immediately cope with the rigorous environment.

Bipolar distribution among invertebrates has been based to a large extent on evidence accumulated from various worm specimens. It is more likely that these are just widely distributed in deep, cold water over broader areas than just the arctic and antarctic. In other words, they are more antitropical than bipolar. The evidence for bipolarity in such forms as molluscs is either lacking or contradictory. And among the sea spiders, for example, there is a marked difference between those at the South Pole and those in the waters of the arctic.

The antarctic, especially, appears to have a large number of endemic invertebrate species, some of which display wide variations with related species as close as New Zealand and Australia.

A distinctive feature of the antarctic is that the continental shelf is sloping, as if it had been bent, and that the dropoff line, often as much as 1500 feet deep, is much lower than in other parts of the world.

*Learning more about **the water world beneath the ice** in polar regions, this diver is exploring an ice formation with a natural or animal-made cave.*

Unique Differences

Fish found in the antarctic generally are found nowhere else. There are occasional wandering individuals, but one group of scientists found that 83 percent of the antarctic fish species considered and recorded in their study were not encountered outside of the region they were examining.

This is not so strange; even though there is an interchange of water between the antarctic and the rest of the world, the flow of currents and temperature distribution in the sea tends to isolate the antarctic waters—in terms of marine life—from other parts of the ocean. Thus it is not surprising that whalers plying the southern waters returned home with tales of bloodless fish having white gills instead of red. This unusual family, the icefish, do indeed have blood in their semitransparent bodies. It's just that their blood appears pale white because it lacks red cells, while there are some nucleated cells that are similar to white blood cells. Ordinarily, red blood cells, with their hemoglobin, are used by vertebrates to carry oxygen through their system. The icefish accomplish this in some other manner, for despite the lack of red cells, their blood is about six percent oxygen by volume, the same as the blood of other fish. A puzzle to be solved, then, is why and how did the ice fish lose their red cells, and if these fish can get along without them, why do other fish have them?

The icefish are not the only fish in the antarctic sea, however. In fact, the most numerous species belong to the order *Nototheniidae*, and include antarctic cods, plunderfish, and dragonfish. The cods, probably the most numerous, are bottom feeders and live on small invertebrates and algae. They are also quite well adapted to living under sea ice. The plunderfish look somewhat like antarctic cods, but have larger heads and, like the icefish, lack scales. The dragonfish are the most distinctive, having an elongated shape and no spiny anterior dorsal fin. Some dragonfish have pointed noses and large canine teeth, and some are almost scaleless.

While these are the most plentiful fish, there are others, sometimes found in deeper waters. These include eelpouts, some true cods roaming from milder waters, flatfish, flounders, and snailfish. These latter, though they are certainly not the most common fish, are widely distributed in the area, in shallow as well as deep water. Snailfish, which are characterized by heavy, gelatinous bodies, are sometimes quite colorful, ranging from an almost transparent white to purple-brown, red-orange, or pink.

*One of the most unusual looking fish in the antarctic waters is the **dragon fish,** whether viewed in profile (opposite, right center) or from above (opposite, top left.) The fish's striped markings (opposite, top right) are an aid in camouflaging his presence.*

*Some varieties of fish are so rarely seen that they prove difficult to identify, such as this **member of the notothenoid family** (opposite bottom) photographed at the bottom of McMurdo Sound.*

*The eerie appearance of the **icefish** (left) was at first attributed to its having no blood. Later it was discovered that these unique fish have a whitish blood coursing through a semi-transparent body.*

Under the Arctic

One of the more unusual findings about life in the Arctic Ocean, the shallowest of all the world's oceans, is that though the fish life is sparse in terms of number of species, there is an abundance of marine invertebrates. The most common fish in the arctic are those that are bottom dwellers, including eelpouts, cod, snailfish, bullheads, sculpins, and flatfishes. The bullheads and armed bullheads have developed perchlike features that make them resemble the polar fish of the antarctic. They have oversized, frogshaped heads and large gill chambers, as well as broad pectoral fins and nonforked tails, and they have poorly developed muscles for lateral movement.

The severity of the environment has an inhibiting effect on fish. Limiting factors include the low temperatures, the destructive ice action near the inshore waters where it is broken up by tidal action, fluctuating salinity due to the freezing and melting of ice, and a lack of large plant life.

An expedition organized under the direction of Dr. Joseph MacInnis of Canada had an observation sphere called "Sub-Igloo," for divers working in Resolute Bay, 600 miles north of the Arctic Circle. Their findings confirmed the disparity between fish species and other marine life—biologists collected fewer than a dozen species of fish and almost 100 different species of invertebrates. One of the biologists, Dr. Alan Emery, described

*Although they may be of different species, genera, and even families, the marine creatures of polar regions often exhibit very similar shape, form, coloration, and patterns of behavior. The specimens shown here are from Antarctica and include a strikingly colorful **jellyfish** (above) and a menacing-looking **isopod** (opposite top). A tagged **starfish** (below) is being eaten by an anemone. Another **anemone** (opposite, bottom right) is devouring a jellyfish, while some **polyps and an isopod** (opposite, bottom left) find themselves in close quarters.*

what he observed: "I found plants and animals more abundant then I ever expected, though compared to the tropics, these waters have much less variety. Without actually diving into the cold depths, we could never have realized how plentiful arctic marine life really is, and yet how painfully slow it grows, moves, and reproduces." Among the more frequent sightings the Sub-Igloo divers made were lion's-mane jellyfish, sometimes with tiny sea fleas, amphipods, hitching a ride on them. There were also numerous ctenophores and nudibranchs, many with partially transparent bodies. Also numerous were tiny sea snails, which dotted the seabed or were feeding on the algae that coated kelp fronds. Flattened armored isopods walked along the bottom, looking very much like extinct trilobites. There were also tiny planktonic crustaceans as well as squids, octopuses, and cetaceans.

111

Below Antarctic Waters

Just as the surface area of the antarctic region can be divided into concentric zones, with the South Pole as the center, the waters around the continent form distinct bands of benthic life. The main characteristics of this benthos is that it completely covers any hard base, that it increases in vitality with depth all the way down to 1500 feet, and that it is more abundant and more exuberant than anywhere else in the world. Not far offshore —in water 50 to 100 feet deep—there is a bare patch caused by large ice crystals, called anchor ice, which form in the bottom during the winter and rise to the surface when the temperatures rise, carrying with them any creatures which may have taken up residence. As a result, during the summer months, only a few mobile creatures like sea stars, ribbon worms, and sea spiders are found in these areas of the polar oceans.

*In antarctic waters there are spider-like **pycnogids** (opposite, top left and bottom) and **starfish** (opposite, top right) which eat **worms** (below).*

Slightly deeper, from 100 to 140 feet, there are many sessile coelenterates, while the next zone contains large starfish, bryozoans, ascidians, isopods, and sponges. The sponges are probably the most common creatures found on the sea floor in the antarctic. There are more sponges found here, displaying their sharp-pointed spicules, than can be found on any tropical seabed.

Corals, which are normally associated with warm waters are found in the antarctic. These are not the colonial, reef-building types, but rather some more singular stony corals and simple corals. Though they are few in number, it does appear the corals are slowly expanding into southern polar waters from more temperate zones.

Among the shelled bivalves, brachiopods are fairly common in areas where there is a rocky substrate or other solid surface to which they can attach themselves. In the absence of rocky bottoms, these brachiopods sometimes attach themselves to the tubes of infaunal polychaetes. In addition to tube worms, there are also wormlike sipunculids and unsegmented echiurids.

There is a greater abundance of molluscs in these waters than was known even recently, and those that are found are generally not much different from those of milder regions slightly to the north. What differences exist are usually due to the fact that the antarctic species live in much deeper waters than those living along South America's coast. Other abyssal organisms found in antarctic waters—although only rarely—are crinoids, sea cucumbers, sea urchins, and sand dollars, and myriads of tiny sea snails.

A remarkable feature of the antarctic shelf is that it is constantly subjected to the destructive action of the thousands of large icebergs drifting with the wind and currents. When they come near shore, the icebergs hit the bottom, pound on it with the swell, and leave a wide trail of disaster behind them.

Chapter IX. The Polar Laboratory

The polar climates, which are as intense as they are harsh, offer examples in the adaptive capabilities of living organisms. While it is not true that evolution has filled every · possible niche with a rare and unique form —there are no terrestrial mammals in Antarctica, for example—there is a tremendous variety of solutions to the problem of living in a restrictive habitat.

The polar regions offer many examples of coping with cold, ice, wind, and water. The animals generally are so specialized that it would be disastrous for them if the climate changed precipitously. The polar bear is a good illustration of this point. It is believed that present-day forms were either developed or highly modified during

---

"The animals generally are so specialized that it would be disastrous for them if the climate changed precipitously."

---

the Great Ice Age, and that the animals were once much more widespread. But since they were equipped to deal with only one kind of environment, they began to die out when the glaciers started to retreat.

Dealing with biology and living creatures, an experimenter or observer can never hope to achieve the same kind of results as a chemist or physicist in a laboratory. But fieldwork can lead to certain generalizations which may have at least limited application. The statements dealing with the relationship between creatures and their environments are referred to as ecological rules. They cover such phenomena as the lighter coloration of mammals and birds in the colder, drier part of their range; the increase in body size of warm-blooded animals as their habitat becomes cooler; and the shortening of wings, legs, tails, and ears as the temperature of the habitats gets lower.

The size and shape of polar animals, as compared with other races of the same species or closely-related species, appear to confirm these rules. And even cold-blooded animals seem to follow these rules on occasion. The largest land animal found in the interior of Antarctica is a wingless fly, *Belgica antarctica*. Even though flies are not warm-blooded, they are affected by cold, so we wonder if their winglessness is favorable to resisting the cold or if it is an answer to the strong and constant buffeting winds which could overpower the tiny fly.

These ecological rules are sometimes applied to man. The traditional comparison is between the short-limbed, rotund Eskimo, living in the arctic, and the lean desert dweller of Asia and Africa. But the problem is that thinness and height in people differ with race. The tallest Caucasians are found along the frost line in northern Europe, while the tallest Mongoloids are found slightly north of this. But among darkskinned Negroids, the tallest people are found in the hot, humid swamps of Ethiopia and the southern Sudan, and in the hot, damp region of Australia.

Biological rules always suffer many exceptions, because we are unable even to imagine the psysiological complexity involved; but in our case, they may prove useful in studying species distribution.

*The extreme cold of Antarctica inhibited the development of land-living mammals. This **young seal** may spend some time on the continent, but is more an aquatic creature than terrestrial beast.*

Distributing Animals

The distribution of animals is limited by such factors as temperature, climate, food supply, shelter, and enemies. In addition, there may be water or land barriers preventing wide ranges. Less important, but still a contributing factor to distribution, is loco-

motion: it is not surprising to find a species of bird with a wider range than a certain species of snail.

Distribution of animals may be discontinous or disconjunctive, such as the related species of marine organisms living at either pole. Most intriguing are those land animals, such as tapirs, which are widely separated. Tapirs

are found only in Asia and in Latin America. How did such closely related animals get so far apart? A similar puzzle is offered by the arthropod *Peripatus,* which is found only in tropical Africa and tropical South America. Continental drift may explain how these creatures have become separated, but the case of the wandering camel remains puzzling. This unique animal is believed to have originated in North America, but there are none there now. Its only close relatives are the llamas of South America and the true camels of Africa.

Why and where certain animals are found involves very many different considerations. It is taken for granted that each species originated in one locality and has either stayed there and spread from that place or has moved entirely away from it. The displacement may have been in reaction to climatic changes. Certain temperate-water codfish were unknown in the North Atlantic a century ago but, because of a warming trend in the ocean, are now commercially fished off Greenland. In other cases, the animals developed such a specialized way of life that they were forced into smaller and more isolated niches. The walrus, with its massive size, its resistance to very low temperatures, and its specialized feeding behavior, has been forced into such a position.

A general law of species diversity can be formulated: in warmer regions there tends to be a greater number of animal species, while in polar regions the number of species is much more limited, but usually with great numbers of individuals per species. The almost overwhelming congregations of penguins in rookeries, the massive amounts of krill, the enormous populations of lemmings all bear this out. The various pinnipeds aggregate in large numbers, and it is hard to imagine how huge their crowds must have been before they suffered near-extinction at the hands of whalers and seal hunters of the recent past.

The flightless **penguin** *is found only in a few areas of the world and these are concentrated on the southern end of the globe. As with most polar animals, there are large numbers of individuals, but only a limited number of species.*

The orca, often referred to as the "killer whale," is found in all the world's oceans, but is most common in polar water, where its great size could be called a special adaptation to cold.

Laws Governing Outlaws

Biological science has several concepts called laws, which really aren't laws at all but rather general statements describing how phenomena naturally occur. Many of the biological laws deal with physiological characteristics and how they change and are modified to meet certain needs presented by the habitat. This selective adaptation takes place through evolution over many generations in a stable environment.

Dollo's law simply states that evolution, except for a rare back-mutation, is a one-way process. Another statement is Cope's law,

formulated by an American paleontologist, which notes that there is a tendency in animals to increase in size until they become extinct. Such laws are not necessarily totally accurate, for they may have been based upon insufficient data or imperfect specimens. Or the laws may be true only in certain circumstances or under certain conditions. They do, however, contain a "grain of truth" that makes them worth noting.

Among the rules dealing more directly with polar areas, Gloger's law states that in the Northern Hemisphere birds and mammals with a north-south distribution tend to display lighter colors in the northern part of their range. Since darker colors are associated with higher humidities, it is not certain whether the humidity or temperature is the determining factor.

In the nineteenth century German biologist Carl Bergmann formulated a law that held that warm-blooded animals tend to be larger in the cool parts of their ranges. The logic behind this, of course, is that the larger the total bulk of the body, the smaller the proportion of surface to volume. This is important in considering that body heat is dissipated most readily through the surface area of an animal, generally the skin.

Bergmann's law helps explain, in part at least, the giant size of whales, which feed mainly in colder waters or in the colder currents of milder environments. The walrus possesses massive size, and the polar bear, though it lives in the most inhospitable of regions, is one of the largest of the mainly meat-eating ursines. In the Southern Hemisphere, the emperor penguins and others ranging the subantarctic are larger than the Magellan penguins, which live in Patagonia and the Falkland Islands. Both are larger than the Humboldt, or Peruvian, penguin living further north.

Bergmann's rule is by no means absolute, for certain temperate-zone animals like raccoons and moles show a decrease in body size as the climate of their range gets cooler.

*The large size of **seals** is one way that these animals have of conserving precious body heat, for the total surface area of the skin is small in relation to the entire bulk of the animal.*

Shrinking Appendages

Coping with cold involves size, body shape, insulation, metabolism, and strategic physiological maneuvers. The larger the animal, the smaller the ratio of surface versus volume, and the smaller the loss of heat per pound of flesh. A spherical shape presents the minimum surface for a given volume: the walrus and the Eskimo are both very plump. Part of the trend toward a spherical shape is the shortening of appendages, as we will see later. Insulation is obtained with blubber or hair, or both. Metabolism is kept at the highest possible level thanks to high central temperature and, in most cases, to abundant and rich food. Finally, the strategic physiological tricks include thermoregulating systems, the control of blood circulation, and, even abandoning body parts temporarily to cold.

The American zoologist Joel A. Allen emphasized the tendency for the appendages of polar inhabitants to become shorter, as compared with races of the same species living in warmer climes. The effect of this is a

lessening of the surfaces from which body heat may be radiated away. Allen's law is evident in a number of polar creatures. The ears of arctic foxes are smaller than the ears of foxes living in milder climates. The arctic hare has ears much shorter than the snowshoe hare which lives further south. The snowshow hare, in turn, has much shorter ears than the antelope jackrabbit, which lives in Arizona.

Seals, whales, and penguins all have extremely short appendages in relation to their body size. Among penguins, there is an even

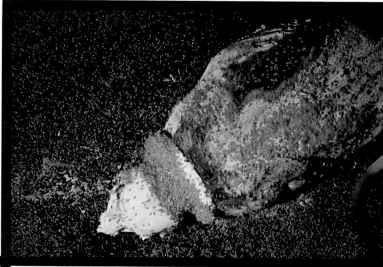

Emperor penguins (left) take advantage of their rounded body shape by "tobogganing" down the slopes, while the life-death-life cycle is shown by the antarctic crustaceans Orchomenella *(above) as they consume a dead* **Weddell seal.**

finer differentiation. The emperor and king penguins are almost indistinguishable in coloration and markings. Yet the emperor, which lives year round on Antarctica or the ice surrounding it, has proportionally shorter flippers, feet, and bill than the king penguin, which lives only a few hundred miles further north. Two other species of penguin display similar characteristics. The Adélie penguin also resides on Antarctica, while the closely related gentoo penguin lives on the subantarctic islands as far north as Gough Island. As expected, the gentoo has a longer bill and bigger feet and flippers than the Adélie.

This tendency described by Allen's rule appears to apply to human beings too. The most northerly living people, the Mongoloid Eskimos, are short, fat, and have proportionally smaller limbs than do the Watusis of Africa. In the south the now-almost-extinct tribe of Yahgan Indians, which numbered in the thousands only a century ago, used to live naked in the snow in the rough climate of Tierra del Fuego. They were small, rounded, and short-limbed.

Chapter X. In Harmony with the Environment

To survive in the arctic means making judicious use of every possible natural resource. The legends and tales of Eskimos are filled not with the exploits of great warriors, but rather with the lessons of learning to cope with the environment and the feats of successful hunters. In the polar regions one must live with nature, not fight it.

It is interesting to note that Eskimos have no one word for snow. This constant part of their life could not be described with so general a term. Rather, the Eskimos describe snow with a variety of terms which refer to its moisture content, texture, or

---

"Garbage dumps are a luxury of affluent and irresponsible societies. For the Eskimos there is nothing to waste."

---

any other physical property that has a practical meaning. Snow is used for so many purposes and comes in such a range of conditions that it would be as difficult for Eskimos to get along with one word for snow as for us to describe plants with just the word flower, for rose and daisy and orchid and tiger lily.

An important principle in making the most of polar life is a conservation of material. Garbage dumps are a "luxury" of affluent and irresponsible societies: the Eskimos have nothing to waste. There is an expression that describes the frugality of Eskimo hunters after a kill: "They make use of everything except the roar." Even human waste products are put to use. Urine, for example was used to ice the runners of a sled—wax was unheard of and the friction of the snow on a cold day could tire the dog team needlessly.

Eskimo or Chukchee, arctic residents were ingenious in their use of animal products. A walrus, for example, provided meat for both people and dogs. Blubber burned in lamps and stoves; hides became boats and floor coverings or were cut into thongs for fishing or harpoon lines. Tusks were carved into hunting implements, the keels of umiak boats, household tools, and religious or decorative figurines. A needle, awl, comb, scoop, knife, buckle, or pair of sun goggles could be made from walrus ivory. In areas where sod houses, rather than skin tents, were built during the summer months, spliced bones were used to support roofs for the houses. The intestines were converted into waterproof rain gear or translucent window coverings, or they were stretched over wooden frames and decorated colorfully to become ceremonial drums.

The Eskimo way of life was developed over centuries of constant contact with the environment. There were many trials and many errors. Occasional innovations came, often through contact with other peoples. Indians to the south developed the use of copper, and the Eskimos traded their furs and hides for metal objects. Then there was extensive trade with the white man. Contact with the white man and his technology led to even greater change.

The white man is now studying the polar world. Instead of time-consuming trial and error based on personal observation, he applies scientific methods in an effort to learn more about the world of ice and snow.

*Part of the non-polluting recreational activities of Eskimo society are **dog-sled races,** such as this one being run outside of Anchorage, Alaska.*

The paddles are used only for emergencies in the **modern Eskimo boats**—*seal hunters now seek out their prey in motor-driven craft.*

In Spite of All

Eskimos had to make a life out of the materials the environment offered. There is little vegetation in the arctic, so the Eskimos did without it. They lived—before the coming of the white sailors and traders—almost entirely on the flesh and fat of animals, whether mammal or fish.

In the early twentieth century the arctic explorer Vilhjalmur Stefansson proved that a non-Eskimo could exist on an all-meat diet in the arctic, and he repeated the experiment

for the medical world in a New York City hospital. He eschewed the more popular steaks and chops for the entrails and head meat of the animals. Stefansson even developed a distinct liking for fishheads, although as a young man he claimed that he could not eat fish without becoming ill. Eventually he came to regard—as the Eskimos did— the fishheads and head meat of mammals as not only the most nutritious of foods, but also the tastiest.

Eating is only part of surviving, and obtaining food is not always easy. To procure their

124

food the Eskimos faced great difficulty. There were no trees to help build fishing boats. What little driftwood there was went into the making of hunting weapons. The boats came in two sizes, and both were made of animal skins: the smaller, covered kayak and the larger, open umiak, used for hunting parties and carrying cargo through open water leads near the shores.

Even housing was a problem, for with no trees and few stones, there was little to find in the way of building materials. The only permanent feature of their environment was ice and snow—so this is what the Eskimos used to build their homes. The design of a snowhouse, or igloo, is as ingenious as it is practical. The small tunnel entrance is inclined slightly upward, so that the warm air inside the house does not escape. The Eskimos learned a long time ago that hot air rises and thus could be neatly and simply trapped within a snow dome by means of a downward-slanting entranceway.

In the warmer months, when even snow isn't available for homes, Eskimos often made tents of animal skins, stitched together with their own sinews in the manner of Indians of the plains far to the south.

*Skin boats are still the main means of water transportation. The larger, **open boats** like this one on Lawrence Island, Alaska are called umiaks.*

Finding the Ends of the Earth

Early exploration of the polar regions was largely a chance occurrence. There was no great lure of gold or other treasure to bring fleets of sailors into the area. There is no evidence that man ever reached Antarctica prior to the nineteenth century, and we think that only the stone age Indians, whose fires inspired Tierra del Fuego's name, reached the subantarctic until the most recent times. Nevertheless, it is amazing that an authentic chart of the world, dated 1531, indicates a continent at the South Pole, looking very much like the real one. Nobody has found an acceptable explanation for it. In the north it is probable that man didn't range into the arctic areas much before 13,000 years ago, deterred by glaciers prior to that.

Historically, the first people known to have visited the arctic were either iron age Norwegians of about 400 A.D. or the Christian missionaries from Ireland who reached Iceland around the third century. Certainly by the ninth and tenth centuries, Vikings were colonizing Iceland and Greenland, and later Vineland. It is probable, however, that the Europeans found some Eskimos already living in the area.

It was not until the search for a Northwest Passage drove sailors into the area that seri-

The icebreaker rides up over the ice and then breaks it with the ship's weight. These ships open channels for smaller craft such as a United States cargo ship and New Zealand research vessel (opposite) or Calypso (below). Icebreakers (above) are also used to push icebergs away from traffic.

ous exploration of the arctic began. John Cabot is traditionally credited with opening the age of arctic exploration in 1496. Shortly afterward came the merchants and traders hoping to capitalize on the wealth of Cathay, but settling for the pelts of animals. Trading firms like the Moscovy Company of London and the Dutch White Sea Trading Company of Amsterdam led the way. Soon explorers, navigators, cartographers, and trappers were moving in, around, and through the ice-dotted sea and offshore islands. The maps of today carry the names of yesterday's adventurers: Hudson, Frobisher, Foxe, and Davis were among the earliest. On the Russian side of the Arctic Circle, the fur traders pushed ever northward in Siberia, but the Asian coast of the Arctic wasn't charted until the mid-eighteenth century, under the direction of Vitus Bering.

All this geographic, adventurous, and mercantile exploration culminated in the wild scramble to the North Pole. The expeditions of Ross, Parry, Simson, Rae, and Franklin set the stage for the hectic race between Frederick Cook and Robert Peary. Though it is debatable whether either man ever reached the North Pole itself, Peary is given credit for doing it in 1909.

Antarctica was explored only from the sea, with no one known to have passed any closer to it than Magellan in 1520, when he sailed through the straits that now bear his name. Maps published shortly after his voyage depict Tierra del Fuego as the northernmost tip of a giant southern continent. Serious exploration of the subantarctic and antarctic regions began with Captain James Cook's circumpolar voyage between 1772 and 1775. It was 50 years later when several rival claims were laid to the first sightings of Antarctica itself. And although John Davis made a reported landing on the mainland in 1821, it was not until 1895 that Henrik J. Bull succeeded in setting a party on the southernmost of continents.

At Sea, Under the Ice

One of the most imposing features of the Arctic Ocean is its year-round covering of ice. As shifting and foreboding a barrier as it is, there is a different world beneath it. The first glimpse of this world came as recently as 1958, when the nuclear submarine *Nautilus* made the first under-ice journey in the Arctic. Not long after that, another U.S. Navy submarine, the *Skate,* not only passed under the North Pole, but became the first vessel to surface at that location. This was accomplished as part of a research mission on conditions of the Arctic Ocean.

A major feature of the ocean floor in the area of the North Pole are pingos. These are large ice intrusions that prevent large-hulled ships —such as the supersized oil tankers—from moving through the water without risking damage. Dr. Robert Dill of the U.S. National Oceanic and Atmospheric Administration describes pingos as stable features of the underwater world, like coral reefs, that have to be charted before the Arctic Ocean can be navigated by deep-hulled vessels. The word

pingo was adopted from the Eskimo, who used it to refer to small mounds of earth or stone forced up by frost action. The mechanisms that cause these ice intrusions are not fully determined.

The lure of arctic waters was first as a potential route to the Orient, the fabled Northwest Passage. Then they were solely a means to the North Pole, over the frozen seawater. The *Nautilus* and *Skate* crews were enthralled at passing under and surfacing, respectively, at the North Pole. The next adventure came in the frigid waters off Corn-

*The **United States submarine** Skate was the first vessel to pass under, and surface at, the North Pole.*

wallis Island, only 125 miles from the magnetic north pole, where a group of divers lived on, and spent long periods under, the ice, obtaining firsthand knowledge of the ocean world of the arctic. The scientific purpose of Operation Sub-Igloo, under the direction of Dr. Joseph B. MacInnis of Toronto, was to study the ability of divers to function in below-freezing waters with a temperature as low as $-28.5°$ F.

To Learn of the World

One of the first systematic attempts to study the polar regions was made by the International Polar Conference, which, during the winter of 1882–83, established a series of scientific stations, eleven in the arctic and four in the antarctic.

In the north this was followed by Fridtjof Nansen's historic journey aboard the *Fram*. This ship, which could both sail and move under steam, was specially designed for use in icy waters. It drifted amid the polar pack ice from 1893 to 1896. *Fram's* hull was rounded on the bottom so that under the pressure of the ice it would rise up rather than be crushed between opposing floes. In 1898 Otto Sverdup used *Fram* for another arctic expedition, and in 1910 Roald Amundsen used the ship in the antarctic in preparing for his successful attempt to be the first to reach the South Pole.

The Russians began establishing drift stations in the 1930s, setting up batteries of men and instruments on large ice floes. After World War II the Soviets expanded their research, using airplanes—which took off and landed on floes—to gather meteorological, oceanographic, and geophysical information. While the Russians chose to use the rather common, but more dangerous, ice floes for their drift stations, the United States sought out the rare, but much thicker and more massive, ice islands for their scientific outposts. The U.S. and Canada didn't begin setting up drift stations until the late 1940s.

Following a fruitful "Polar Year" in 1932–33, the International Geophysical Year of 1957–58, involving 30,000 scientists and technicians from 66 nations, brought an increase in scientific activity. Many of these were stationed in the polar regions and their efforts not only capped, but also initiated intense research efforts.

Earlier antarctic study included many unscientific attempts by adventurers and explorers who collected odd specimens and natural rarities, such as penguins and their eggs. There were some serious attempts at obtaining geographic data and some more important efforts aimed at studying unusual deposits found in the ice-free areas of the antarctic mountain ranges. But for the most part, scientific endeavor at the South Pole was dominated by publicity and commercialism, overlaid with the land claims and counterclaims made on the territory by sovereign nations. The International Geophysical Year, however, imposed some order on the chaos. The major efforts were directed toward meteorology and upper atmosphere physics, best studied in polar regions; the earth's magnetic field; the depth of the antarctic ice cap and the nature of the land beneath it; and Antarctica's effect on the world's weather.

Antarctic research stations (below) are sparsely furnished and lack many amenities of home, but sometimes they offer the chance to observe the unique such as on January 29, 1967 when a rocket was blasted into the ionosphere (above) not far from the French base at Dumont d'Urville. The 25-foot-long "Dragon" was only the first of a series of rockets which the French scientists sent aloft under the watchful eyes of penguins in Adélie Land, at the south magnetic pole, about 930 miles from the geographic South Pole.

Chapter XI. To Plunder or to Manage the Poles

Man, at least so it seems, is incapable of leaving a place the way he found it, whether it is outer space, the moon, or Antarctica. The mark of man's presence is often an absence—the disappearance of blue whales in familiar waters or the lack of fur seals on certain islands. Elsewhere man leaves something behind, a physical reminder.

Antarctica is a continent recently dedicated to science, for the 500 winter residents and the 5000 summertime guests are engaged in learned work or aid it indirectly by providing support services. Yet even with such noble purpose, Antarctica is being ruined. The most staggering problem is waste disposal. Garbage dumps are difficult to dig in land that is permanently frozen. Containers, packaging, and odds and ends would pose problems even if they were all biodegradable, which they in fact certainly are not. Decay and return to organic use of the waste product would be a solution to the problem,

> "Containers, packaging, and odds and ends would pose problems even if they were all biodegradable, which they in fact are not."

but in frigid waters the processes of recycling in this manner are so slow that garbage would accumulate faster than it could be accommodated. Petroleum-based plastics are virtually indestructible under such circumstances. Even incineration is no great solution; not only does burning pollute the air with foreign chemicals, but there is always residue. Incineration may buy time, but even here there is a catch. The scavenger skuas of the South Pole, which have learned to live off man's trash piles, have been killed by incinerators, for they had never encountered anything like fire before and were roasted by the flames while looking for a meal in the mess.

Power plants, no matter what kind of fuels they use, are another disaster at the poles. Fossil fuels pollute the air, and nuclear plants can superheat the water and cause a problem of radioactive waste disposal.

Man is not always conscious of his own plundering. Certainly he has no reason to denude an island or territory of its birds or grass. But ships bring rats; sailors bring dogs and cats; settlers bring sheep and goats and rabbit. Overnight an environment has changed. Birds—like the blue petrel—that through generations of evolution developed the ability to nest in burrows for protection from flying predators suddenly become acquainted with a rat's maze-ranging talents. Strains of grass and plants hardy enough to withstand polar climates begin to yield and finally die out under the persistent feeding of herbivorous hares and rabbits.

Even in the arctic, where man has had a much longer association, assaults are being made. The polar routes used by jet airplanes are paths of pollutants. Oil exploration on the continental shelf of North America promises the wellhead leaks, seepage, and tanker accidents familiar to the North Sea and Gulf of Mexico.

What to do, what to do? The hope is that man will somehow be able not only to define the problems, but also divine the solutions.

The Alaska pipeline will bring oil from the North Slope to southern Alaskan ports. Ecological problems may arise in the process and the merits of oil versus life may have to be decided.

Dead Bears Tell No Tales

Polar bears are the most curious of animals, perhaps always wondering who is intruding on their domain. Humans traversing the arctic regions are often followed for several miles by investigating bears. If a camp is abandoned or even temporarily vacated, polar bears poke around, chewing every manner of intriguing object and even biting through tins to sample the contents.

Bears are not particularly fond of men as meals, much preferring seals and walruses. Thus, if a party of humans has seal meat in its camp, the bear often hunts out the seal meat without disturbing much else, except some peace of mind. It is not that the bears are brave, or foolhardy, but they enter the camps because they have not yet learned to fear men. The bear has no enemies on land, and it has found that humans are as often a source of food as they are a threat to life.

Its curiosity and sense of well-being have worked against the bear. Hunting parties in search of trophies—the great white hide, a mounted head, or a "stuffed" animal—have to do little stalking. As often as hunters have found their quarry, bears have sought out their killers. Even before polar bear shooting was a sport, Eskimos used the victim's own curiosity against it. If a group of hunters could not draw a polar bear close enough for their spears or arrows, perhaps because the bear shied away from a large hunting party, one of the Eskimos would attract the bear's attention by shouting and waving his arms while his companions lay in hiding. Unless this particular bear had learned from past experience, it will come to investigate, usually getting close enough to be set upon by the whole group.

It takes no great skill to shoot a polar bear, for once it is trapped by dogs, a bear usually takes a stand with its back to a hummock, or on top of it. From this position it swipes its paws at the dogs or the man holding a lance. But with a gun it is no contest. The bear may even be hit a number of times before it takes another course of action in an effort to escape, usually jumping into water if it is nearby. The hunter, on the other hand, risks little, since the rifle allows the shameful luxury and safety of distance.

Early arctic explorers report seeing great numbers of bears, which were promptly killed out of fear or ignorance. The white skins became popular curiosities in Europe.

Later the whaling men began to kill the bears, and reports of the Dundee whalers from 1905 to 1909 indicate that they killed more than 3000 walruses and more than 1000 polar bears east of Greenland.

The hunting of polar bears—and their main source of food, the seals—continued unchecked until the 1960s. The hunting was done by natives who used only one out of every five bears for themselves; by commercial hunters with airborne scouting parties, high-powered rifles, and telescopic sights; and by wealthy "sportsmen" who paid several thousand dollars for the privilege of taking potshots at the white beast.

The total number of polar bears in the world is estimated variously between 6000 and 20,000, with about 11 per cent, or one in nine, being killed each year. However many polar bears there actually are, it is certainly many times fewer than existed when man first began to hunt them.

*Because of declining numbers, **polar bears,** like this one, are being protected and moved about as part of the International Polar Bear Program.*

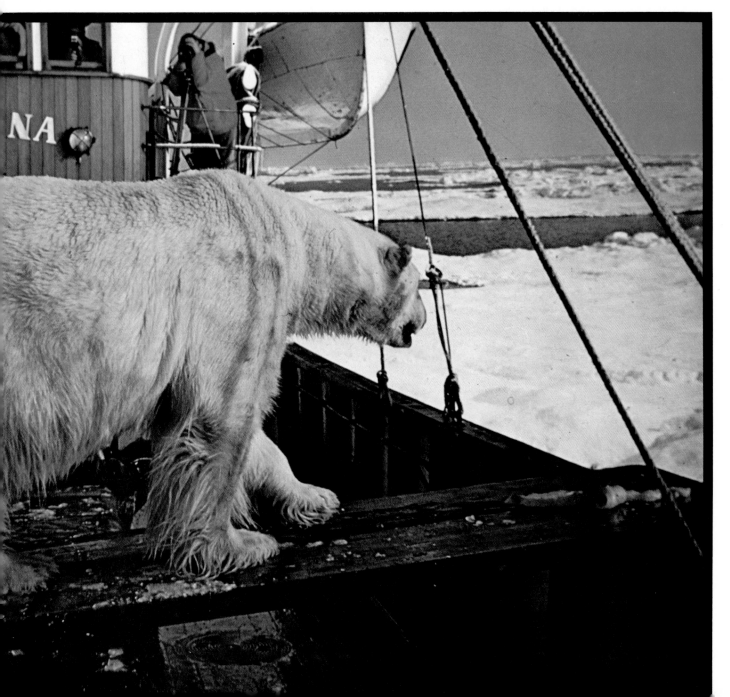

Manipulating the Eskimo

Native hunters used to kill for need. Polar bears provided hides for warm clothing, boots, and gloves; the stringy meat fed valuable dogs; bones, hair, teeth, everything was put to use. Seals, walruses, caribou also yielded 100 percent useful products. The polar peoples didn't stockpile caribou hides. No one tried to corner the market on walrus tusks. If nothing was needed, nothing was killed. It was that simple.

Things changed when the white man came. Hungry whalers who were too busy or inex-

perienced to hunt for themselves began trading tea and tobacco for caribou meat. The natives learned to kill for other than their own need. Arctic explorer Vilhjalmur Stefansson, writing of his stay at Point Shingle in the MacKenzie River delta in the winter of 1906–07, described the scene: "Some years earlier caribou had been in the habit of coming down to the coast frequently, but the Eskimos told me that dur-

*Life in the arctic was difficult, but **polar bears** managed well enough until man came along to collect specimens for zoos. In captivity, bears become neurotic and have a tendency to put on weight.*

ing the last few years, they had been so much hunted by the natives in the employ of the whalers that none were to be expected now north of the mountains, which were 20 or 30 miles inland."

After the Eskimos began killing supplies for the whaling ships and sealers, they found that there was more money to be made in seal, caribou, and polar bear hides. The killing of a polar bear in winter, with its long outer coat and woolly underlayer, became more advantageous in dollars-and-cents terms for the natives than killing only what was needed for their own use. Eskimos went from living off the land to a cash economy. The desire for polar bear skins and live specimens, especially cubs for zoos, has helped deplete the world population and encouraged circumvention of international agreements against the exploitation of these and other animals. Technology, too, has played a part; animals are no match for man, with his spotter planes, helicopters, snowmobiles, and compact camping equipment.

There may be good reason to use snowmobiles, such as in rescue, police, or transportation services, but too often they become vehicles of destruction when used by hunters to decimate caribou herds.

The Big Wipe-Out

Modern whaling, especially as still practiced by the Japanese and Russians, combines a number of small but very fast "super-catcher" boats which do the actual hunting and killing, and a huge factory ship, where the flensing and processing is done. The processing—converting the blubber to oil and making cosmetics, pet food, or jelly of the blood and bones—is aided by steam winches, power saws to hack apart the body, and auxiliary ships built to rigid specifications for speed and efficiency. Even radar is used, for after a whale-catcher makes a kill, it doesn't bother to haul it to the factory ship. Rather it inflates the carcass with compressed air so it will float, unless it has a naturally buoyant sperm whale body, and implants a metal detector that the factory ship's radar can detect.

These whaling techniques have developed only recently, and the whole industry is less than a hundred years old in the antarctic. It wasn't until the beginning of this century

*Whaling has changed considerably since its techniques were immortalized in Herman Melville's "Moby Dick." The **harpoon** is still the basic kill weapon (above). Evidence of past whaling is strewn about on antarctic islands, where we reconstructed the **90-foot long skeleton** of a blue whale (below).*

that the first shore station was established on the island of South Georgia.

Preceding the whalers were the seal hunters, who had discovered the antarctic's rich supply of fur seals and elephant seals at least as early as the 1700s. The kills were as fantastic as they were profitable. One ship, the *Aspasia* of New York, reported having obtained 57,000 seal skins in 1800. A quarter of a century later, British sealer Capt. James Weddell estimated that, from South Georgia alone, the total number of skins taken was not less than 1,200,000 and the amount of oil obtained from the blubbery elephant seals amounted to at least 20,000 tons. If there is one thing that whalermen and sealers have in common, it is that neither group took steps to assure that there would be future populations of their prey. Animals were killed indiscriminately—males, females, young and old. In the land-based stations, when fuel for the melting pots ran low, penguin fat was burned. And when the elephant seals were killed off, the penguins themselves found their way into the pots.

The extent of devastation is almost unbelievable. Macquerie Island was discovered in 1810, and within 10 years the hundreds of thousands of fur seals that had called it home had been killed, every last one of them. It took another decade for the elephant seal population to be wiped out. Next came the king penguins, which numbered in the millions. Perhaps that was too many, for a few survived. When the king penguins were so few as to become unprofitable to kill, the royal penguins succeeded them as a source of oil. Every royal penguin that has known freedom has known Macquarie Island as its birthplace. They, too, were somehow able to survive the death of more than a million of their number in a few years' time. Soon there was nothing left for the sealers to do but leave. They did, but not for good, for as soon as the elephant seals had repopulated Macquarie from other islands, the sealers returned and almost wiped the elephants from the map again. They were not successful. Their only lasting "tribute" was an end of the fur seal on Macquarie Island.

What to Expect

Hope is an ephemeral thing. It can be easily shattered by the unscrupulous, the ignorant, the distrusting, and the economically deprived. The very nations that found it profitable to exploit the polar regions—England, the United States, Russia, and Japan—are the countries that must be relied upon to protect these areas.

There are accords and agreements of every international stripe, restricting the numbers and types of animals that can be killed and specifying those that are to be spared. But for every agreement, there is more than one individual willing to break or ignore it. The International Whaling Commission is a case in point. This organization was founded with noble intent, but it is often frustrated with procedures and paperwork. In the end, it is only as effective as its member nations want it to be. And no nation is forced to become a member of the commission.

Perhaps wasteful killing will stop. There may even be a time when no animals will need be killed, for fashions changes, substitute products can be found, zoos may learn to breed rare animals in captivity.

An era of hope came with the signing of the Antarctic Treaty of 1959. The signatories were Argentina, Australia, Belgium, Chile, France, Great Britain, New Zealand, Norway, South Africa, the Soviet Union, and the United States. They agreed to waive any territorial claim for 30 years, to ban nuclear explosions from the continent, provide for inspection of each other's scientific stations, and to take unresolvable disputes to the International Court of Justice. But even more in line with the spirit of cooperation that prevailed, the treaty stipulated that "it is in the interest of all mankind that Antarctica shall continue forever to be used exclusively for peaceful purposes and shall not become the

scene or object of international discord." Thus, a continent became dedicated to science and peaceful purposes for all peoples based on the open exchange of information between nations.

But there is another, more pernicious, danger to the animals and the polar environments.

It is, again, man. Not man the hunter, but man the consumer. Modern, technological people waste as much as they use, maybe more, of electrical power and fossil fuel by obsolete equipment, overproduction, and uneconomical usage. Slowly, as man contaminates the water, poisons the air, he is killing

*South of Adelaide Island, the **Calypso crew** organized balloon ascents to count animal populations.*

himself. To reverse the trend before it is too late, all human beings have not only to change their approach to nature, but probably their very way of life.

Index

CHARTS AND ILLUSTRATIONS:

PHOTO CREDITS: